Thin Film Diamond

Thin Film Diamond

Edited by A. Lettington and J.W. Steeds

Published by Chapman & Hall for The Royal Society

The Royal Society

CHAPMAN & HALL
London · Glasgow · New York · Tokyo · Melbourne · Madras

Published by Chapman & Hall, 2–6 Boundary Row, London SE1 8HN

Chapman & Hall, 2–6 Boundary Row, London SE1 8HN, UK

Blackie Academic & Professional, Wester Cleddens Road, Bishopbriggs, Glasgow G64 2NZ, UK

Chapman & Hall Inc., One Penn Plaza, 41st Floor, New York NY 10119, USA

Chapman & Hall Japan, Thomson Publishing Japan, Hirakawacho Nemoto Building, 6F, 1-7-11 Hirakawa-cho, Chiyoda-ku, Tokyo 102, Japan

Chapman & Hall Australia, Thomas Nelson Australia, 102 Dodds Street, South Melbourne, Victoria 3205, Australia

Chapman & Hall India, R. Seshadri, 32 Second Main Road, CIT East, Madras 600 035, India

First edition 1994

Typeset by Thomson Press, India
Printed in Great Britain by the University Press, Cambridge

ISBN 0 412 49630 5

A catalogue record for this book is available from the British Library

Library of Congress Cataloging-in-Publication data

Thin film diamonds / edited by A. Lettington and J.W. Steeds. – 1st ed.
p. cm.
Includes index.
ISBN 0-412-49630-5
1. Diamonds, Artificial. 2. Diamond thin films. I. Lettington, Alan H. II. Steeds, J. W. III. Royal Society (Great Britain)
TP873.5.D5T47 1994
666.88–dc20 93-33950
CIP

Printed on permanent acid-free text paper, manufactured in accordance with the proposed ANSI/NISO Z39.48-1992 and ANSI/NISO Z39.48-1984 (Permanence of Paper).

Contents

Contributors

T. Ando National Institute for Research in Inorganic Materials, 1-1 Namiki, Tsukuba, Ibaraki 305, Japan

John C. Angus Department of Chemical Engineering, Case Western Reserve University, Cleveland, Ohio 44106, USA

T. R. Anthony GE Research & Development Center, P.O. Box 8, Schenectady, New York 12309, USA

Alberto Argoitia Department of Chemical Engineering, Case Western Reserve University, Cleveland, Ohio 44106, USA

Peter K. Bachmann Philips Research Laboratories, Solid Films and Deposition Technologies, P.O. Box 1980, D-5100 Aachen, Germany

James E. Butler Code 6174, Naval Research Laboratory, Washington, DC 20375-5000, USA

C.D. Clark J. J. Thomson Physical Laboratory, University of Reading, Whiteknights, Reading RG6 2AF, UK

Alan T. Collins Wheatstone Physics Laboratory, King's College London, Strand, London WC2R 2LS, UK

C. B. Dickerson J. J. Thomson Physical Laboratory, University of Reading, Whiteknights, Reading RG6 2AF, UK

Z. Feng Computer Mechanics Laboratory, University of California at Berkeley, USA

J. E. Field Cavendish Laboratory, Madingley Road, Cambridge CB3 0H3, UK

H. Fujita Central Research Laboratory, Onoda Cement Co. Ltd, 2-2-1 Ishikawa, Sakura, Chiba 285, Japan

Roy Gat Department of Chemical Engineering, Case Western Reserve University, Cleveland, Ohio 44106, USA

R. Haubner Technical University Vienna, Institute for Chemical Technology of Inorganic Materials, Getreidemarkt 6/161, A-1060 Vienna, Austria

M. Kamo National Institute for Research in Inorganic Materials, 1-1 Namiki, Tsukuba, Ibaraki 305, Japan

Alan H. Lettington J. J. Thomson Physical Laboratory, University of Reading, Whiteknights, Reading RG6 2AF, UK

Zhidan Li Department of Chemical Engineering, Case Western Reserve University, Cleveland, Ohio 44106, USA

B. Lux Technical University Vienna, Institute for Chemical Technology of Inorganic Materials, Getreidemarkt 6/161, A-1060 Vienna, Austria

E. Nicholson Cavendish Laboratory, Madingley Road, Cambridge CB3 0HE, UK

J. Robertson National Power Labs, Leatherhead, Surrey KT22 7SE, UK

Y. Sato National Institute for Research in Inorganic Materials, 1-1 Namiki, Tsukuba, Ibaraki 305, Japan

M. Seal Sigillum B.V., P.O. Box 7129, Amsterdam 1007 JC, The Netherlands

C. R. Seward Cavendish Laboratory, Madingley Road, Cambridge CB3 0HE, UK

Mahendra Sunkara Department of Chemical Engineering, Case Western Reserve University, Cleveland, Ohio 44106, USA

T. Tanaka National Institute for Research in Inorganic Materials, 1-1 Namiki, Tsukuba, Ibaraki 305, Japan

Long Wang Department of Chemical Engineering, Case Western Reserve University, Cleveland, Ohio 44106, USA

Yaxin Wang Department of Chemical Engineering, Case Western Reserve University, Cleveland, Ohio 44106, USA

Richard L. Woodin Norton Diamond Film, Northboro, Massachusetts 01532-1545, USA

Preface

This volume contains a selection of invited review papers presented at a Royal Society Discussion meeting on Thin Film Diamond held in London on 15 and 16 July 1992.

The topic of low pressure synthesis has attracted world wide interest and become increasingly active in recent years due to the possible use of diamond films in commercial applications.

Until recently commercial diamond synthesis was almost entirely by the high pressure high temperature technique in which diamond is precipitated as an equilibrium phase from a carbon-containing liquid metal catalyst. In this way crystals may be formed up to 10 mm or so in size. The metastable low pressure techniques cannot compete in cost but can be used to fabricate large area wafers or predetermined shapes not possible by other means.

Most of the low pressure techniques stem from the work of Eversole which was first reported in 1962. He exposed a hot diamond substrate alternately to a hydrocarbon gas, which deposited a mixture of diamond and graphite, and then to hydrogen, which preferentially etched away the graphite. In later developments these two stages have been combined to form a continuous process and differ only in the way the etchant is generated. In this volume an historical overview of these low pressure growth techniques and a description of diamond and crystal morphology is given by John Angus and his co-authors in their paper on the chemical vapour deposition of diamond. James Butler and Richard Woodin discuss the kinetics and gas phase chemistry involved in thin film growth and Peter Bachmann reviews the current deposition techniques. He also summarizes the results of various deposition conditions and shows that diamond growth is possible only in a narrow range of gas compositions. Y. Sato and co-workers report on the local epitaxial growth of diamond on nickel substrates.

Other papers discuss the electronic, optical, thermal and mechanical properties of thin diamond films. The relationship between diamond and diamond-like carbon (DLC) is now better understood and this volume also contains papers on the electronic structure, deposition techniques and applications of DLC films.

The final paper in this volume discusses the various thermal and optical infrared and X-ray applications of diamond thin films. They have also been used in cutting and grinding and as the active element in semiconducting devices.

The subject is advancing rapidly, with many patents being taken out each year. It is nevertheless an opportune time to review the field since there is much fundamental and applied research that needs to be undertaken before this topic can realize its full potential. It is hoped that this volume will provide a valuable overview of thin film diamond for years to come.

A. H. LETTINGTON
J. W. STEEDS

1

Chemical vapour deposition of diamond

BY JOHN C. ANGUS, ALBERTO ARGOITIA, ROY GAT, ZHIDAN LI, MAHENDRA SUNKARA, LONG WANG AND YAXIN WANG

Growth of diamond at conditions where it is the metastable phase can be achieved by various chemical vapour deposition methods. Atomic hydrogen plays a major role in mediating rates and in maintaining a proper surface for growth. Low molecular weight hydrocarbon species (e.g. CH_3 and C_2H_x) are believed to be responsible for extension of the diamond lattice, but complete understanding of attachment mechanisms has not yet been achieved. The nucleation of diamond crystals directly from the gas phase can proceed through a graphitic intermediate. Once formed, the growth rate of diamond crystals is enhanced by the influence of stacking errors. Many of the commonly observed morphologies, e.g. hexagonal platelets and (apparent) decahedral and icosahedral crystals, can be explained by the influence of simple stacking errors on growth rates. *In situ* measurements of growth rates as a function of hydrocarbon concentration show that the mechanism for diamond growth is complex and may involve surface adsorption processes in rate limiting steps. The transport régime in diamond deposition reactors varies widely. In the hot-filament and microwave reactors, which operate from 20 to 100 Torr (1 Torr $\approx$ 133 Pa), the transport of mass and energy is dominated by molecular diffusion. In the atmospheric pressure combustion and plasma methods, transport is dominated by convection. *In situ* measurements of H atom recombination rates in hot-filament reactors show that, under many commonly used process conditions, transport of atomic hydrogen to the growing surface is diffusion limited and H atom recombination is a major contributor to energy transport.

---

1. Introduction

Diamond synthesis by chemical vapour deposition (CVD), at pressures and temperatures where diamond is metastable with respect to graphite, was first achieved by William G. Eversole of the Union Carbide Corporation (Eversole 1962). Unpublished reports show that Eversole achieved growth of new diamond during the period 26 November 1952 to 7 January 1953 (A. D. Kiffer). This is just before 15 February 1953, the date of the first high-pressure–high-temperature synthesis of diamond by Liander & Lundblad (1960) at Allemana Svenska Eliktriska Aktiebolaget (ASEA), and well ahead of the synthesis by General Electric in 1954.

Efforts to grow diamond at low pressures arose independently of each other and without knowledge of the Eversole work in both the Soviet Union and the United States (Spitsyn & Deryagin 1956; Angus 1961). The first published papers from these groups did not appear until much later (Angus *et al.* 1968; Deryagin *et al.* 1968). The use of atomic hydrogen to etch graphitic deposits (Angus *et al.* 1971) and its use during growth to permit high rate nucleation and growth of diamond on non-

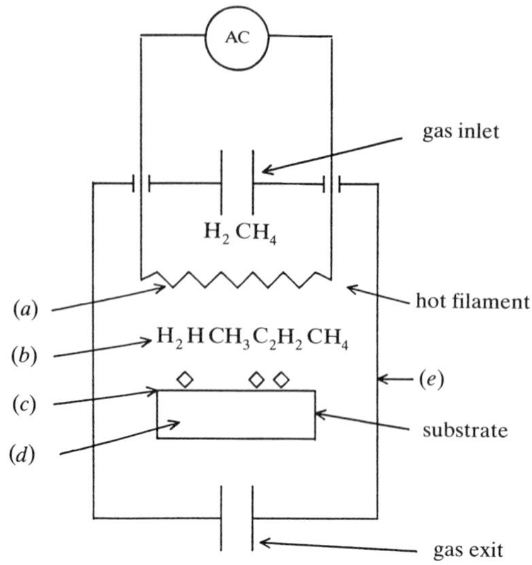

Figure 1. Schematic diagram of hot-filament chemical vapour deposition reactor (permission of *A. Rev. Mater. Sci.*). Reaction numbers refer to table 1. (*a*) Reaction 1; (*b*) gas phase reactions 2–6; (*c*) surface reactions 7–9; (*d*) solid state diffusion, carbide formation; (*e*) wall reactions.

diamond substrates (Deryagin *et al.* 1976) followed. The current intense level of interest can be traced to the Japanese group at the National Institute for Research in Inorganic Materials (NIRIM) in Tsukuba, Japan. They revealed details of several methods for the CVD of diamond, including the widely used hot-filament and microwave plasma assisted techniques (Matsumoto *et al.* 1982*a*, *b*; Kamo *et al.* 1983; Matsui *et al.* 1983).

Details of the history of diamond synthesis have been extensively reported elsewhere and will not be repeated here (Davies 1984; DeVries 1987; Badzian & DeVries 1988; Angus 1989, 1990).

Diamond has long been known to be thermodynamically stable with respect to graphite only at high pressure (Leipunski 1939; Berman & Simon 1955). This fact was erroneously interpreted by many workers to mean that diamond could never be synthesized at low pressures, where graphite is the stable form of carbon. A notable exception to this view was Bridgman (1955).

Diamond is, in fact, only slightly unstable with respect to graphite. At 298 K and 1 atm† pressure, the difference in free energy between diamond and graphite is only 2.900 kJ mol^{-1} (approximately 0.03 eV per atom), which is only on the order of kT. Furthermore, there is a very large activation barrier inhibiting the transformation of diamond to graphite. Of perhaps more significance is the fact that graphite (2.26 g cm^{-3}) is less dense than diamond (3.51 g cm^{-3}). Therefore as solid carbon precipitates out of a supersaturated carbon gas, the 'first' phase encountered is graphite, not diamond. This is an example of Volmer's rule, which states that the least dense solid phase is the first to crystallize from a supersaturated fluid phase. Once formed, graphite will not spontaneously transform further into diamond because it is the stable phase and because of the high activation barrier between the two phases. However, atomic hydrogen changes the relative energies of small graphitic and diamond nuclei and provides a means for circumventing the large activation barrier. This will be discussed in more detail in §3 below.

In CVD of diamond from C/H and C/H/O gas mixtures the gas phase is decomposed to atomic hydrogen, molecular fragments, free radicals and sometimes ions. The most commonly used methods are various types of plasma discharges, heated filaments and combustion. Diamond is formed from these complex gas phase reaction mixtures on substrates held at nominal temperatures from 700 °C to 1000 °C. A simple schematic of a hot-filament reactor is shown in figure 1. Although details of the complex reaction and transport processes are still the subject of much research, a general understanding of the reactor environment and the principal reactions and energy flows has been achieved.

2. Diamond growth

Some of the principal reactions of importance are summarized in table 1 with an estimate of their standard enthalpy and free energy changes. Examination of table 1 gives a clear picture of the basic energetics of the entire diamond deposition process. The principal function of the hot filament, or equivalently the plasma discharge, is to provide large amounts of atomic hydrogen through the decomposition of molecular hydrogen (reaction 1). This reaction has a very positive enthalpy change.

The atomic hydrogen, once formed, undergoes several spontaneous, highly exothermic, reactions. It can react with hydrocarbons in the gas phase, abstracting hydrogen to form methyl radicals (reaction 2). Recombination of atomic hydrogen to form molecular hydrogen (reaction 3) is also possible. However, since this is a three body reaction, its rate is slow at low reactor pressures and often can be ignored despite the favourable free energy change. Methyl radical destruction can take place by recombination with atomic hydrogen (reaction 4) or by diffusion out of the reaction zone to the walls of the reactor. There are gas phase reaction paths to higher molecular weight species, e.g. reaction 5. A spectrum of C_2H_x species can be formed by subsequent hydrogen atom abstraction reactions of the general type shown in reaction 6. These can lead to still higher molecular weight compounds by subsequent reactions not shown in table 1.

Atomic hydrogen will hydrogenate a bare diamond surface (reaction 7). Hydrogenated diamond surfaces are less prone to surface reconstruction than bare surfaces. On (111) surfaces, for example, bonded hydrogen helps maintain the bulk terminated diamond surface structure (Lander & Morrison 1964). Hydrogen can also abstract hydrogen from a hydrogen covered surface (reaction 8). This reaction is thermodynamically favoured because of the strong H–H bond. At typical substrate temperatures the fractional coverage is dominated by the competition between reactions 7 and 8. To first order, the steady state concentration of free radical sites, f_s, is independent of the atomic hydrogen concentration and is approximately given by the competition between reactions 7 and 8

$$f_s = k_8/(k_7 + k_8), \tag{1}$$

where k_7 and k_8 are the first order rate constants for reactions 7 and 8. f_s can be estimated using kinetic constants for analogous gas phase reactions. At temperatures

Table 1. *Standard enthalpy and free energy changes of some important reactions during diamond growth*

(1 kcal = 4.184 kJ.)

reaction		$T/(K)$	$\Delta H^0/(\text{kcal mol}^{-1})$	$\Delta G^0/(\text{kcal mol}^{-1})$[a]
	on filament			
1	$H_2 \rightarrow 2H$	2500	+109	+37[b]
	in gas phase			
2	$H + CH_4 \rightarrow CH_3 + H_2$	1800	*ca.* 0	−11
3	$H + H + M \rightarrow H_2 + M$	1800	−108	−57
4	$CH_3 + H + M \rightarrow CH_4 + M$	1800	−108	−45
5	$CH_3 + CH_3 + M \rightarrow C_2H_6 + M$	1800	−86[c]	−26[c]
6	$C_2H_x + H \rightarrow C_2H_{x-1} + H_2$	1800	small	small
	on substrate (S)			
7	$H + S \rightarrow S\text{–}H$	1200	−94	−68
8	$S\text{–}H + H \rightarrow S + H_2$	1200	−13	−6
9	$CH_3 + S \rightarrow S\text{–}CH_3$	1200	−81	−47

[a] Rossini & Jessup (1938), Stull *et al.* (1969) and Stull & Prophet (1971).
[b] At 20 Torr and 2500 K, the equilibrium mole fraction of atomic H is 0.16.
[c] Molecular mechanics estimate.

of 1000 and 1750 K, f_s was estimated to be 0.12 and 0.37 respectively (Kuczmarski 1992). These sites are where free radicals such as CH_3 (reaction 9) or acetylenic species, C_2H_x, can add to the surface. Finally, the H atom recombination on the surface contributes substantially to the energy flux delivered to the substrate. The recombination could be direct through reaction 3 where the surface plays the role of the third body or it could result as the net reaction from the two step process of reactions 7 and 8.

Much attention has focused on identification of the gas phase 'growth species' that are responsible for the addition of carbon to the diamond surface. Methyl radical, CH_3, and acetylene, C_2H_2, are believed to be the most likely candidates. Recent evidence, for example the isotope labelling studies of Chu *et al.* (1991) indicate that the methyl radical is the primary source of added carbon. The state of understanding has been the subject of an excellent recent review by Celii & Butler (1991) in which they summarized the extensive studies of Tsuda, Frenklach, Harris, Hauge and others. A recent molecular dynamics study by Garrison *et al.* (1992) of growth on the diamond (001) surface is also of great interest.

Recent microbalance studies of diamond growth kinetics show a complex behaviour. In a hot-filament reactor at 20 Torr†, the growth rate of diamond was first order in methane from 0.1% to 1.0% CH_4 (Wang *et al.* 1992). At high methane concentrations, the reaction tended to zero order. For the two-carbon source gases, C_2H_2, C_2H_4 and C_2H_6, the reaction was approximately one half order for concentrations from 0.3% to 1.0% hydrocarbon. Below a concentration of 0.3% hydrocarbon, the two-carbon gases showed a first order rate (see figure 2). The data are in agreement with a much earlier study (Chauhan *et al.* 1976) in which first-order kinetics were found for CH_4 and half order for C_2H_4. These data can be rationalized by mechanisms that involve single-carbon atom species in rate limiting steps. The zero-order dependence on CH_4 at high concentrations may arise from adsorbed intermediates.

† 1 Torr ≈ 133 Pa.

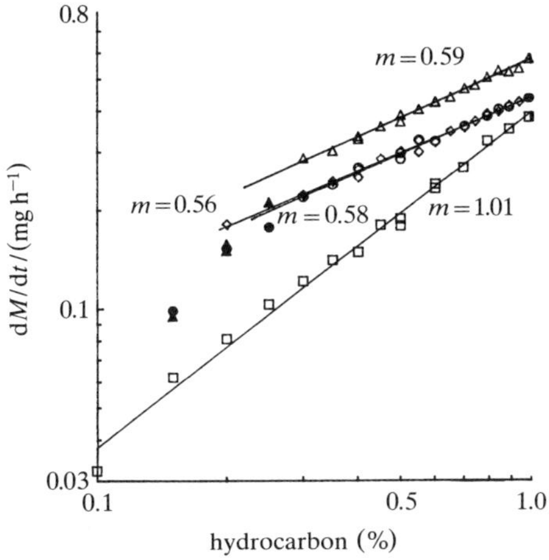

Figure 2. Log reaction rate against mole hydrocarbon (%) for various C_1 and C_2 gases. □, CH_4; ◇, C_2H_2; ○, C_2H_4; △, C_2H_6.

3. Diamond nucleation

The spontaneous nucleation of new diamond crystals under conditions where graphite is the stable carbon phase has been difficult to rationalize. Matsumoto & Matsui (1983) proposed, on the basis of symmetry arguments, that hydrocarbon cage compounds might serve as diamond precursors. It was proposed that a more likely precursor for diamond nucleation would be graphitic intermediates, which are subsequently hydrogenated by atomic hydrogen to saturated structures that can act as sites for diamond growth (Angus *et al.* 1988; Sunkara *et al.* 1990). Recent experimental studies support this latter view. Belton & Schmieg (1990) studied the nature of carbon bonding at different stages of nucleation on platinum substrates using X-ray photoelectron spectroscopy. They found that a carbon phase with graphitic bonding first formed, followed by a hydrogenated carbon phase and finally diamond. Microbalance studies of diamond nucleation and growth on platinum (Wang *et al.* 1992) show an initial induction period in which an oriented graphite deposit forms. Subsequently, this deposit disappears and the final deposit contains only polycrystalline diamond. A plot of the mass against time data and the observed Raman signals are shown in figure 3.

The nature of diamond nucleation can be probed by using graphite flakes as seed crystals. If the graphite is acting as a true nucleation site and not simply as an extra source of carbon, one would expect to find an orientational relation between the diamond and the original graphite substrate. Initial experiments on diamond growth on graphite seed crystals showed that the (111) diamond plane was parallel to the basal (0001) plane of the graphite (Angus *et al.* 1991). Subsequent experiments show that in addition to (111) diamond ∥ (0001) graphite, one often has a directional orientation within the planes, i.e. $[1\bar{1}0]$ diamond ∥ $[11\bar{2}0]$ graphite (Li *et al.* 1992). This relation means that the puckered six-membered rings in the diamond (111) planes retain the same orientation as the flat six-membered rings in the graphite basal (0001) plane. A transmission electron micrograph of one of these oriented

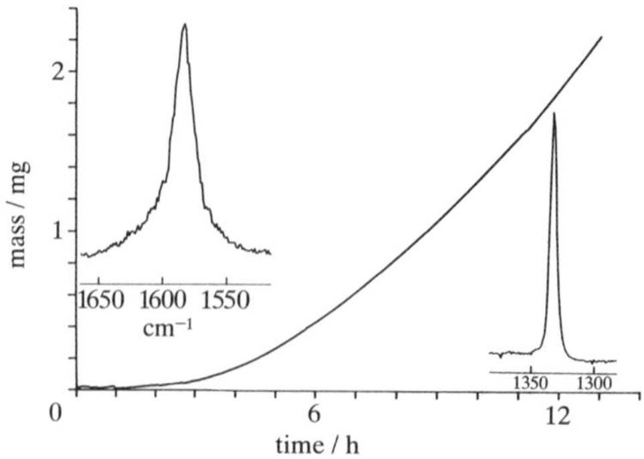

Figure 3. Mass of diamond against time in a microbalance hot-filament reactor. Inset at upper left shows Raman signal of graphite obtained during initial period (peak at 1581.8 cm^{-1}). Inset at lower right shows Raman signal of diamond obtained during steady state growth period (peak at 1333.1 cm^{-1}).

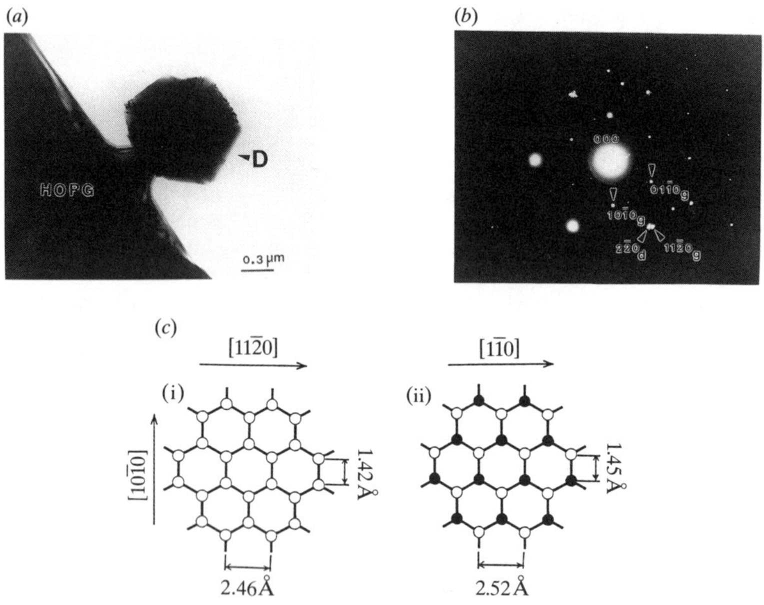

Figure 4. (*a*) Transmission electron micrograph of diamond nucleated on edge of highly oriented pyrolytic graphite. View is along normal to (0001) basal plane of graphite and the (111) plane of diamond. (*b*) Electron diffraction pattern from diamond and graphite shown in (*a*). (*c*) Orientation of (0001) graphite plane (i) and (111) diamond plane (ii).

diamond crystals on graphite, the corresponding diffraction pattern and the geometric relationship between the two structures are shown in figure 4.

The energetics of the conversion of graphitic, sp^2 bonded precursors can be examined by simple thermochemical calculations. Badziag *et al.* (1990) pointed out that hydrogen terminated 'diamonds' less than 3 nm in diameter have a lower

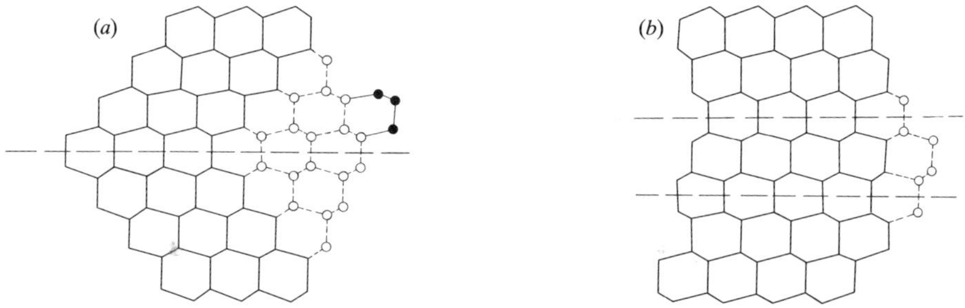

Figure 5. (*a*) ⟨110⟩ projection of diamond lattice containing one stacking error. Open circles and dashed lines show crystal filling out to smooth {111} surfaces with a convex corner. Three-atom nucleus shown on {111} surface with filled circles (permission *J. Mater. Res.*); (*b*) ⟨110⟩ projection of diamond lattice containing two stacking errors (extrinsic stacking fault). Open circles show crystal filling out to give a re-entrant (concave) corner. Two-atom nucleus shown with closed circles. Further growth can occur without the necessity of a three-atom nucleation event.

energy than hydrogen terminated graphite nuclei with the same number of carbon atoms. This means that in an environment rich in atomic hydrogen, the sp^3, tetrahedrally coordinated nuclei are in fact energetically favoured over the sp^2, trigonally coordinated nuclei. This work was subsequently criticized by Stein (1990), who pointed out that the correct parameter to consider is the free energy change for the appropriate reaction. However, Stein neglected to account for the fact that the active reagent under diamond growing conditions is atomic hydrogen, H, not molecular hydrogen, H_2. The enthalpy and free energy changes for the sequential hydrogenation of graphite to naphthalene and decalin by atomic hydrogen are:

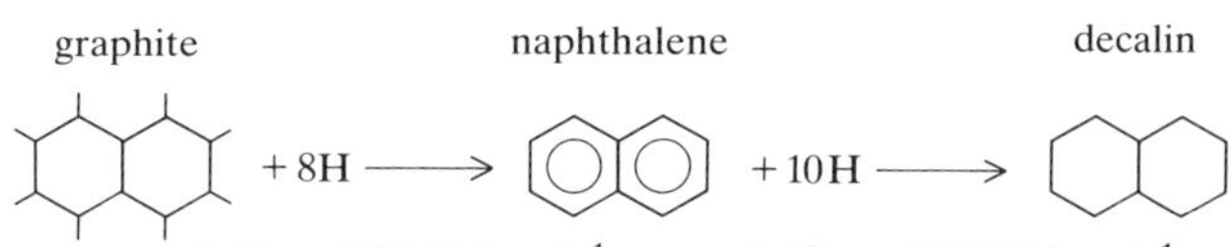

$\Delta G^\circ = -904.6\ \text{kJ mol}^{-1}$ $\quad\Delta G^\circ = -1338.3\ \text{kJ mol}^{-1}$

$\Delta G \leqslant -356.4\ \text{kJ mol}^{-1}$ $\quad\Delta G \leqslant -653.1\ \text{kJ mol}^{-1}$

(ΔG was calculated from nominal reaction conditions of 0.01 atomic fraction of hydrogen, pressure of 20 Torr and equal activities of naphthalene, decalin and graphite. Data were taken from Stull *et al.* (1969).) The estimated free energy changes at reaction conditions are strongly negative for these model reactions, which shows that graphitic nuclei can indeed be converted into hydrogen saturated structures similar to diamond. Molecular orbital studies of the hydrogenation of single graphite sheets also support this conclusion (Angus *et al.* 1991; Mehandru *et al.* 1992). In addition, cyclopropane and cyclohexane are found as products when graphite is reacted with atomic hydrogen (Rye 1977). Other workers have also used thermodynamic methods to show that hydrogen stabilizes diamond surfaces (Sommer *et al.* 1989; Piekarcyk *et al.* 1989; Yarbrough *et al.* 1990; Harris *et al.* 1991).

The nucleation sequence may start with the formation of high molecular graphitic and polynuclear aromatic hydrocarbons (PAH) by the sequential polymerization of acetylene (Frenklach *et al.* 1985). These high molecular weight materials are sufficiently non-volatile so that they remain on the substrate until they become hydrogenated by atomic hydrogen, forming the saturated edge structure that is

(*a*)

(*b*)

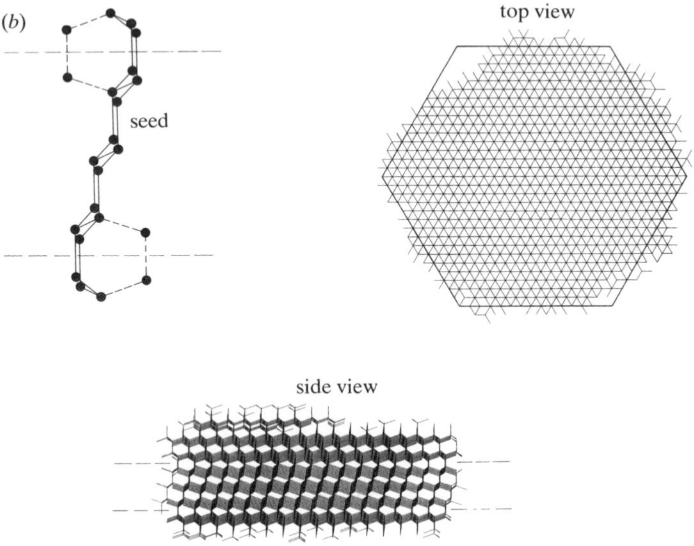

Figure 6. (*a*) Hexagonal diamond platelet and fully faceted three-dimensional diamond crystal. Re-entrant grooves are visible along the side faces of the hexagonal platelet and on the fully faceted crystal (marked with Gs). (*b*) Seed nucleus and top and side views of resulting hexagonal platelet grown by Monte Carlo simulation. Stacking errors are shown by dashed lines.

attractive for diamond nucleation. The atomic hydrogen plays a multiple role in this process. By terminating the dangling surface bonds it stabilizes the tetrahedrally coordinated, sp^3 nuclei with respect to the trigonally coordinated, sp^2 nuclei. It also serves as a reactive solvent which permits the conversion of graphitic nuclei into diamond nuclei, thereby circumventing the large activation barrier separating graphite from diamond. The mechanism is consistent with the careful observations of Lux, who showed that induction times for nucleation are shortest on those metals that can achieve a supersaturation of carbon on the surface the most rapidly (Joffreau *et al.* 1988; Lux & Haubner 1991).

Other methods of diamond nucleation are also possible. For example, growth

Table 2. *Relation of observed morphology to types of stacking errors during growth of {111} faceted crystals*

type of error	morphology
two stacking errors on parallel (111) planes (intrinsic or extrinsic stacking fault or micro twin)	hexagonal platelet
three stacking errors on parallel (111) planes	truncated hexagonal platelet
two stacking errors on non-parallel (111) planes	decahedral (pseudo five-fold symmetry)
three stacking errors on non-parallel (111) planes	icosahedral
single stacking error	triangular (macle)

on diamond debris left by scratching the substrate surface with diamond powder is commonly done (Iijima 1991).

The observed morphologies of vapour-grown diamond crystals range include: spherical clusters of diamond microcrystals, cubes, cubo-octahedrons, octahedrons and flat hexagonal platelets. Complex, multiply twinned forms (e.g. decahedrons and icosahedrons) are also observed. Twinned clusters, with many re-entrant surfaces, are perhaps the most common form, especially when in a regime that gives octahedral {111} faceting.

Defects that arise during growth can mediate further growth of the crystal. It appears that re-entrant corners play a major role in enhancing diamond growth rates. The re-entrants arise from the intersection of {111} twin bands or stacking faults with the surface. The re-entrant corners are favourable sites for nucleation of new layers. This effect has been known for many years in other contexts and was exploited for the rapid growth of silicon web from the melt. Presumably, the enhanced nucleation rate arises because only two atoms are required to form a stable nucleus at the re-entrant corner rather than three atoms that are required on a smooth {111} surface. This is shown schematically in figure 5. A recent discussion of the nucleation at re-entrant corners has been given by Tiller (1991).

Many complex morphologies can arise from the enhancement of growth rates by the re-entrant corners arising from multiple stacking errors (Sunkara 1992; Angus *et al.* 1991, 1992). A Monte Carlo computer program was used in which atoms were added to seed 'crystals' with different types of stacking errors (Sunkara 1992). A computer 'grown' hexagonal platelet and a scanning electron micrograph of a hexagonal diamond platelet are shown in figure 6. This morphology arises from the action of two parallel stacking errors. In the model shown in figure 6, the two stacking errors are separated by two layers of correct stacking. The growth rate within the plane of this microtwin is much greater than the growth rate normal to the plane because of the re-entrant corner on the periphery of the crystal. The result is a flat platelet of hexagonal shape and low aspect ratio. Hexagonal diamond platelets have been observed by several workers (Everson & Tamor 1991; Angus *et al.* 1991, 1992) and may reflect the nature of the precursor that leads to diamond nucleation. In a much earlier study, hexagonally shaped platelets of silver bromide were attributed to the mechanism described above (Berriman & Herz 1957).

Many other morphologies commonly observed in vapour-grown diamond can also be explained by the interaction of various combinations of stacking errors. These are summarized in table 2.

Table 3. *Comparison of mass and energy transport in diamond reactors*

type of reactor	Pe_{mass}	Pe_{thermal}
hot filaments	6×10^{-4}	8×10^{-4}
microwave	5×10^{-4}	7×10^{-3}
plasma torch	60	300
combustion	12	7

(Note that

$$Pe_{\text{mass}} = \frac{\langle u\rangle L}{D}\left[\frac{\text{mass flux by convection}}{\text{mass flux by diffusion}}\right],$$

$$Pe_{\text{thermal}} = \frac{\langle u\rangle L}{\alpha}\left[\frac{\text{thermal flux by convection}}{\text{thermal flux by diffusion}}\right],$$

where $\langle u\rangle$ is the average convective velocity, L is a characteristic length, D is the diffusion coefficient and α is the thermal diffusivity.)

4. Reactor environment

Much less attention has been paid to the transport processes within the reactor than to the attachment kinetics at the diamond surface. This is somewhat surprising since there is growing evidence that diamond growth rates may be limited by transport processes (Angus *et al.* 1989; Rau & Picht 1992). A complete understanding of diamond deposition must include a simultaneous solution of the energy and mass transport equations together with a detailed chemical mechanism. Simplified reactor models have been described (Debroy *et al.* 1990; Goodwin & Gavilett 1991; Kuczmarski *et al.* 1991). In the absence of a detailed reactor model, much of interest can be learned from an analysis of existing growth rate data.

A comparison of the transport of species and energy between the various types of reactors has been made (Angus *et al.* 1989, 1991). In the low pressure processes (e.g. microwave and hot-filament-assisted deposition) the transport is entirely dominated by molecular diffusion. Convection plays virtually no role. On the other hand, in the atmospheric pressure plasma torch and in atmospheric combustion, the transport is dominated by convection. An estimate of the magnitude of these effects can be obtained by computing average Peclet numbers. These results are shown in table 3. The growth rates in the atmospheric pressure, high gas velocity processes are much higher than in the low pressure processes. This is suggestive that the growth rate is limited by the rate of transport of species rather than by the attachment kinetics.

Similarly, the thermal Peclet numbers indicate that in typical hot-filament and microwave reactors, heat transport by conduction is much more important than heat transport by forced convection. The opposite is true in the plasma torch and combustion reactors.

The importance of natural convection in hot-filament reactors can be estimated from the Grashof number, Gr,

$$Gr = g\beta L^3 \Delta T/\nu^2, \tag{2}$$

where g is the acceleration of gravity, β the volumetric expansion coefficient, L a characteristic length, ΔT a temperature difference and ν the kinematic viscosity. The Grashof number is a measure of the relative magnitude of buoyancy forces to viscous forces. For a hot-filament reactor containing primarily H_2 at 20 Torr, a ΔT of 1400 K and a characteristic (substrate to filament) length of 0.8 cm, we find $Gr \approx 0.7$, indicating that free convection is marginal at 20 Torr. This estimate indicates that in typical hot-filament reactors the flow is marginally stable to free convection and to roll cells. This conclusion is supported by direct measurements of heat transfer in a dual-filament diamond deposition reactor (Gat 1992; Gat & Angus 1992) and modelling studies (Kuczmarski *et al.* 1992; Kuczmarski 1992). The ability to use microbalances within the hot-filament reactor is further testimony to the unimportance of convection at these conditions.

Recent studies have shown that a significant amount of energy is transported by the decomposition of H_2 and subsequent recombination of H atoms (Yarbrough *et al.* 1992; Gat 1992; Gat & Angus 1992). Quantitative measurements of the energy released by the recombination of atomic H were made in a novel dual-filament reactor (Gat 1992). The power required to maintain a tungsten substrate wire at a fixed temperature was measured as a function of various process variables. By measuring the power to the substrate filament with an adjacent hot-filament on and off, the contribution of the atomic H recombination to the total energy transport could be quantitatively determined. The results show that in a typical hot-filament reactor, with the gas phase primarily hydrogen at 20 Torr, the energy transport by H atom recombination is approximately equal to that transferred by conduction. Furthermore, the contribution of convection is small at 20 Torr. The measurements of H atom recombination rate permits the determination of the net H atom diffusion flux from the hot-filament to the substrate. Under typical growth conditions there are approximately 10^4 H atom recombination events for each carbon atom added to the diamond. The large H atom recombination rate on the diamond surface leads to large gradients in H atom concentration in the gas phase. For one-dimensional transport between the hot-filament and the substrate the ratio of concentrations of atomic hydrogen, C_{H}, at the substrate and the hot-filament is given by

$$\frac{C_{\mathrm{H,s}}}{C_{\mathrm{H,hf}}} = \frac{1}{1+(2kT/\pi m)^{\frac{1}{2}}(\gamma l/D)}, \tag{3}$$

where D is the average diffusion coefficient of H, γ is the recombination coefficient, l is the filament to substrate distance, and m the mass of a hydrogen atom. For typical hot-filament conditions ($D = 0.175\ \mathrm{m^2\,s^{-1}}$, $l = 0.01$ m, $T = 1000$ K and $\gamma = 0.5$), $C_{\mathrm{H,s}}/C_{\mathrm{H,hf}} \approx 0.1$. This is in general agreement with the results of Hsu (1991) and with recent reactor modelling studies in which the energy and species transport equations were solved for simple hot-filament geometries (Kuczmarski 1992). Despite the very large diffusion coefficient of H, the transport of H to the surface is partly diffusion limited.

5. Summary

A general understanding of the factors permitting diamond growth at low pressures has been achieved, although detailed molecular mechanisms are not known with certainty. Stacking errors formed during growth enhance growth rates and strongly influence the final crystal morphology. Nucleation of diamond can occur through graphitic intermediates or by growth on diamond debris left from scratching. In many processes the growth rates of diamond appear to be limited by transport processes rather than by attachment kinetics.

The support of a National Science Foundation Materials Research Group grant is gratefully acknowledged.

References

Angus, J. C., Will, H. A. & Stanko, W. S. 1968 Growth of diamond seed crystals by vapor deposition. *J. appl. Phys.* **39**, 2915–2922.

Angus, J. C., Gardner, N. C., Poferl, D. J., Chauhan, S. P., Dyble, T. J. & Sung, P. 1971 *Sin. Almazy* **3**, 38–40.

Angus, J. C., Hoffman, R. W. & Schmidt, P. H. 1988 Studies of amorphous hydrogenated diamondlike hydrocarbons and crystalline diamond. In *Science and technology of new diamond* (ed. S. Saito, O. Fukunaga & M. Yoshikawa), pp. 9–16. Tokyo: KTK/Terra.

Angus, J. C. 1989 History and current status of diamond growth at metastable conditions. In *Proc. First Int. Symp. on Diamond and Diamondlike Films*, pp. 1–13. Pennington, New Jersey: Electrochemical Society.

Angus, J. C., Buck, F. A., Sunkara, M., Groth, T. F., Hayman, C. C. & Gat, R. 1989 Diamond growth at low pressures. *MRS Bulletin* October, 38–47.

Angus, J. C., Li, Z., Sunkara, M., Gat, R., Anderson, A. B., Mehandru, S. P. & Geis, M. W. 1991*a* Nucleation and growth processes in chemical vapor deposition of diamonds. *Electrochem. Soc. Symp. Series.* Pennington, New Jersey: Electrochemical Society.

Angus, J. C., Wang, Y. & Sunkara, M. 1991*b* Metastable growth of diamond and diamondlike phases. *A. Rev. Mater. Sci.* **21**, 221–248.

Angus, J. C. 1991 Innovations in the chemical vapor deposition of diamond: perceptions of a participant. In *Japanese/American technological innovation* (ed. W. David Kingery), pp. 136–142. New York: Elsevier.

Angus, J. C., Sunkara, M., Sahaida, S. R. & Glass, J. T. 1992 Twinning and faceting in the early stages of diamond growth by chemical vapor deposition. *J. Mater. Res.* **7**, 3001–3009.

Badziag, P., Verwoerd, W. S., Ellis, W. P. & Greiner, N. R. 1990 Nanometre-sized diamonds are more stable than graphite. *Nature, Lond.* **343**, 244–245.

Badzian, A. R. & DeVries, R. C. 1988 Crystallization of diamond from the gas phase: part 1. *Mater. Res. Soc. Bull.* **23**, 385–400.

Belton, D. N. & Schmieg, S. J. 1990 States of surface carbon during diamond growth on Pt. *Surf. Sci.* **233**, 131–140.

Berman, R. & Simon, F. 1955 On the graphite-diamond equilibrium. *Z. Elektrochem.* **59**, 333–338.

Berriman, R. W. & Herz, R. H. 1957 Twinning and the tabular growth of silver bromide crystals. *Nature, Lond.* **180**, 293–294.

Bridgman, P. W. 1955 Synthetic diamonds. *Scient. Am.* **193**, 42–46.

Celii, F. G. & Butler, J. E. 1991 Diamond chemical vapor deposition. *A. Rev. Phys. Chem.* **42**, 643–684.

Chauhan, S. P., Angus, J. C. & Gardner, N. C. 1976 Kinetics of carbon deposition on diamond powder. *J. appl. Phys.* **47**, 4746–4754.

Chu, C. J., D'Evelyn, M. P. & Hauge, R. H. 1991 Mechanism of diamond growth by chemical vapour deposition on diamond (100), (111) & (110) surfaces: carbon-13 studies. *J. appl. Phys.* **70**, 1695–1705.

Davies, G. 1984 *Diamond*. Bristol: Adam Hilger Ltd.

Debroy, T., Tankala, K., Yarbrough, W. A. & Messier, R. 1990 Role of heat transfer and fluid flow in the chemical vapor deposition of diamond. *J. appl. Phys.* **68**, 2424–2432.

Deryagin, B. V., Spitsyn, B. V., Builov, L. L., Klochkov, A. A., Gurodetski, A. E. & Smolyaninov, A. V. 1976 Synthesis of diamond on non-diamond substrates. *Dokl. Akad. Nauk SSSR* **231**, 333–335.

Deryagin, B. V., Fedoseev, D. V., Spitsyn, B. V., Lukyanovich, D. V., Ryabov, B. V. & Lavrentev, A. V. 1968 Filamentary diamond crystals. *J. Cryst. Growth* **2**, 380–384.

DeVries, R. C. 1987 Synthesis of diamond under metastable conditions. *A. Rev. Mater. Sci.* **17**, 161–176.

Eversole, W. G. 1962 Synthesis of diamond. U.S. Patents 3,030,187 and 3,003,188.

Everson, M. P. & Tamor, M. A. 1991 Studies of nucleation and growth of boron-doped diamond microcrystals by scanning tunneling microscopy. *J. Vac. Sci. Technol.* B **9**, 1570–1576.

Frenklach, M. 1991 Molecular processes in diamond formation. In *Proc. Second Int. Symp. on Diamond Materials* (ed. A. J. Purdes, J. C. Angus, R. F. Davis, B. M. Meyerson, K. E. Spear & M. Yoder), vol. 91–98, pp. 145–153. Pennington, New Jersey: Electrochemical Society.

Frenklach, M., Clary, D. W., Gardiner, W. C. & Stein, S. E. 1985 Detailed kinetic modeling of soot formation in shock-tube pyrolysis acetylene. In *Proc. 20th Int. Symp. on Combustion*, pp. 887–901. Pittsburgh: The Combustion Institute.

Garrison, B. J., Dawnkaski, E. J., Srivastava, D. & Brenner, D. W. 1992 Molecular dynamics simulations of a dimer opening on a diamond (001) (2×1) surface. *Science, Wash.* **255**, 835–842.

Gat, R. 1992 Hot filament assisted deposition of diamond films. Ph.D. thesis, Case Western Reserve University, Cleveland, Ohio.

Gat, R. & Angus, J. C. 1992 Energy transport in hot-filament assisted deposition of diamond films. *J. appl. Phys.* (In the press.)

Goodwin, D. G. & Gavillet, G. G. 1991 Numerical modeling of filament-assisted diamond growth. In *Proc. 2nd Int. Conf. on New Diamond Science and Technology* (ed. R. Messier & J. T. Glass), pp. 335–340. Pittsburgh: Material Research Society.

Harris, S. J. & Wiener, A. M. 1985 Chemical kinetics of soot particle growth. *A. Rev. Phys. Chem.* **36**, 31–52.

Harris, S. J., Belton, D. N. & Blint, R. J. 1991 Thermochemistry on the hydrogenated diamond (111) surface. *J. appl. Phys.* **70**, 2654–2659.

Iijima, S., Aikawa, Y. & Baba, K. 1991 Growth of diamond particles in chemical vapor deposition. *J. Mater. Res.* **6**, 1491–1497.

Joffreau, P. O., Haubner, R. & Lux, B. 1988 Low pressure diamond growth on refractory metals. In *Proc. MRS Spring meeting.* Pittsburgh: Material Research Society.

Kamo, M., Sato, S., Matsumoto, S. & Setaka, N. 1983 Diamond synthesis from gas phase microwave plasma. *J. Cryst. Growth* **62**, 642–644.

Kuczmarski, M. A. 1992 Modelling of chemical vapor deposition reactors for silicon carbide and diamond growth. Ph.D. thesis, Case Western Reserve University, Cleveland, Ohio.

Kuczmarski, M. A., Washlock, P. A. & Angus, J. C. 1991 Computer simulation of a hot-filament CVD reactor for diamond deposition. In *Applications of diamond films and related materials* (Mater. Sci. Monographs) (ed. Y. Tzeng, M. Yoshikawa, M. Murakawa & A. Feldman), vol. 73, pp. 591–596. Amsterdam: Elsevier.

Lander, J. J. & Morrison, J. 1964 Low energy electron diffraction study of the (111) diamond surface. *Surf. Sci.* **2**, 241–246.

Leipunski, O. I. 1939 Synthetic diamonds. *Usp. Khim.* **8**, 1519–1534.

Li, Z., Wang, L., Suzuki, T., Argoitia, A., Pirouz, P. & Angus, J. 1992 Orientation relationship between chemical vapour deposited diamond and graphite substrates. *J. appl. Phys.* (In the press.)

Liander, H. & Lundblad, E. 1960 Some observations on the synthesis of diamonds. *Ark. Kemi.* **16**, 139–149.

Lux, B. & Haubner, R. 1991 Nucleation and growth of wear-resistant diamond coatings. In *Proc. Electrochemical Society Spring meeting.* Pennington, New Jersey: Electrochemical Society.

Matsui, Y., Matsumoto, S. & Setaka, N. 1983 TEM-electron energy loss spectroscope study of the diamond particles prepared by chemical vapor deposition from methane. *J. Mater. Sci. Lett.* **2**, 532–534.

Matsumoto, S., Sato, Y., Kamo, M. & Setaka, N. 1982*a* Vapor deposition of diamond particles from methane. *Jap. J. appl. Phys.* **2**, L183–L185.

Matsumoto, S., Sato, Y., Tsutsumi, M & Setaka, N. 1982*b* Growth of diamond particles from methane-hydrogen gas. *J. Mater. Sci.* **17**, 3106–3112.

Mehandru, S. P., Anderson, A. B. & Angus, J. C. 1992 Hydrogen binding and diffusion in diamond. *J. Mater. Res.* **7**, 689–695.

Piekarcyk, W., Roy, R. & Messier, R. 1989 Application of thermodynamics to the examination of the diamond CVD process from hydrocarbon-hydrogen. *J. Cryst. Growth* **98**, 765–776.

Rau, H. & Picht, F. 1992 Rate limitation in low pressure diamond growth. *J. Mater. Res.* **7**, 934–939.

Rossini, F. D. & Jessup, R. S. 1938 Heat and free energy of formation of carbon dioxide, and the transition between graphite and diamond. *J. Res. NBS* **21**, 491–513.

Rye, R. R. 1977 Reaction of thermal atomic hydrogen with carbon. *Surf. Sci.* **69**, 653–667.

Spitsyn, B. V. & Deryagin, B. V. 1956 Process for growing diamond grains. Author's patent certificate dated July 10, 1956; U.S.S.R. patent 339,134, May 5, 1980.

Stein, S. E. 1990 Diamond and graphite precursors. *Nature, Lond.* **346**, 517.

Sommer, M., Mui, K. & Smith, F. W. 1989 Thermodynamic analysis of the chemical vapor deposition of diamond films. *Solid State Commun.* **69**, 775–778.

Stull, D. R., Westrum, E. F. & Sinke, G. C. 1969 *The chemical thermodynamics of organic compounds*. New York: John Wiley.

Stull, D. R. & Prophet, H. 1971 *JANAF thermochemical tables*, 2nd edn. Washington, D.C.: National Bureau of Standards.

Sunkara, M. 1992 Monte Carlo simulation of diamond nucleation and growth. Ph.D. thesis, Case Western Reserve University, Cleveland, Ohio, U.S.A.

Sunkara, M., Angus, J. C., Hayman, C. C. & Buck, F. A. 1990 Nucleation of diamond crystals. *Carbon* **28**, 745–746.

Tiller, W. A. 1991 *The science of crystallization: microscopic interfacial phenomena*. New York: Cambridge University Press.

Wang, Y., Evans, E. & Angus, J. C. 1992 Diamond growth kinetics. *J. appl. Phys.* (Submitted.)

Yarbrough, W. A., Inspektor, A. & Messier, R. 1990 The chemical vapour deposition of diamond. In *Properties and characterization of amorphous carbon films* (ed. J. J. Pouch & S. A. Alterovich), pp. 151–174. Aedermannsdorf, Switzerland: Trans. Tech.

2

Thin film diamond growth mechanisms

BY JAMES E. BUTLER AND RICHARD L. WOODIN

The principal chemical mechanisms relevant to the growth of diamond from gaseous hydrogen and hydrocarbon species are presented. The kinetic processes occurring during the activation and transport of the gaseous species to the growing surface are described, with the key processes being the generation and subsequent reactions of gaseous atomic hydrogen. The structure, composition, and dynamics of the growing surface are discussed. A simple, non-stereospecific model of the surface growth process is presented which reveals most of the general characteristics of the growth process, such as the H atom flux dependence of growth rate and quality. A detailed model of growth at the (110) surface sites from single carbon reactants then follows, which highlights the key role of gaseous atomic hydrogen abstractions of hydrogen from the surface. The extension of this understanding to chemistries containing oxygen and halogen species is indicated.

1. Introduction

The chemical vapour deposition (CVD) of diamond from gases at pressures between 10 mTorr† and 1 atm‡ is now an established technology (Celii & Butler 1991). There are diverse implementations of this technology, using hot filaments, plasmas, combustion flames, and other techniques to drive the deposition process. The reactants are typically a small amount of methane (or other hydrocarbon) in hydrogen, but variations have included carbon monoxide, trace amounts of oxygen, combustible gas mixes such as oxygen–acetylene, halogens, and halocarbons (Bachmann *et al.* 1991). A diamond CVD process is schematically summarized in figure 1. Gaseous reactants, typically 0.5% methane in hydrogen at *ca.* 50 Torr, flow into the reactor and gaseous reactions are initiated by the hot filament or plasma. The reactants, products, and reactive species are transported throughout the reactor by diffusion and convection. For higher pressure processes, a stagnation layer is formed just above the deposition surface through which the reacting species must diffuse. At lower pressures, this near surface chemical environment may differ from that of the bulk gas because of surface chemical reactions. On the surface, adsorption, diffusion, reaction and desorption of various species occurs leading to the nucleation of diamond particles, suppression of graphitic (or sp^2) carbon, and ultimately the growth of a continuous diamond film. This is a complex process with many variations, and in which many specific mechanisms occur. The intent of this paper is to present a discussion of the principal processes occurring during the CVD

† 1 Torr ≈ 133 Pa. ‡ 1 atm ≈ 10^5 Pa.

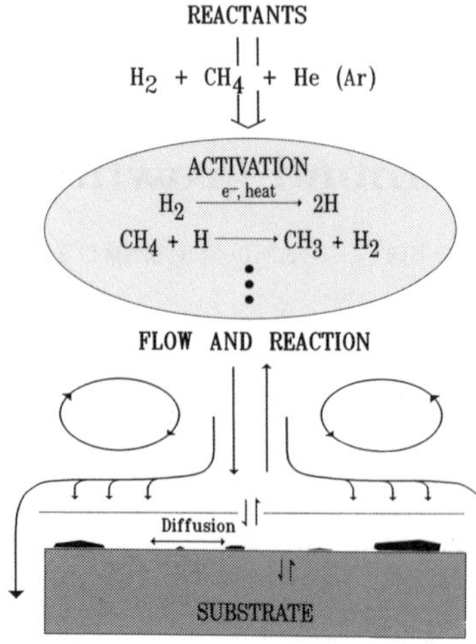

Figure 1. A schematic showing the principal elements in the complex diamond CVD process: flow of reactants into the reactor, activation of the reactants by thermal and plasma processes, reaction and transport of the species to the growing surface, and surface chemical processes depositing diamond and other forms of carbon.

of diamond, from the initial gaseous reactions to the incorporation of carbon into the diamond lattice. The selection of these principal processes is judgmental and is based on our experience and a wealth of literature on the diagnostics, characterization, and implementation of diamond CVD processes.

This discussion is constrained to diamond CVD processes using hydrogen and dilute concentrations of hydrocarbon species (less than 1% (by volume)). This environment contains all of the important features of diamond CVD and rationalizes the empirical growth diagram of Bachmann *et al.* (1991) for C–H–O environments when the chemically stable CO species is ignored. Exceptions and refinements to the following arguments exist and should be considered after the basic processes are understood. The gaseous hydrocarbons considered are limited to one and two carbon atom species, consistent with the major species observed (Celii *et al.* 1988; Celii & Butler 1989, 1991, 1992; Harris *et al.* 1988; Wu *et al.* 1990; Hsu 1991*a*, *b*). The gaseous activation process, whether by plasmas or a hot filament, is considered as a source of atomic hydrogen, as observed by laser diagnostics and optical emission measurements (Celii & Butler 1989; Mucha *et al.* 1989; Chen *et al.* 1992). The surface growth processes focus on the homo-epitaxial growth on diamond in the (110) direction, the fastest growing direction observed in most studies (Wild *et al.* 1990; Specht *et al.* 1991; Chu *et al.* 1991, 1992).

2. Gas phase chemistry and transport

Initiation of the gaseous chemistry is dominated by dissociation of 1–40% of the molecular hydrogen (which is over 99% of the reactant mixture) into atomic hydrogen, depending on the environment. This can be by thermal dissociation on a hot filament or by electron impact dissociation in a plasma. The subsequent gaseous

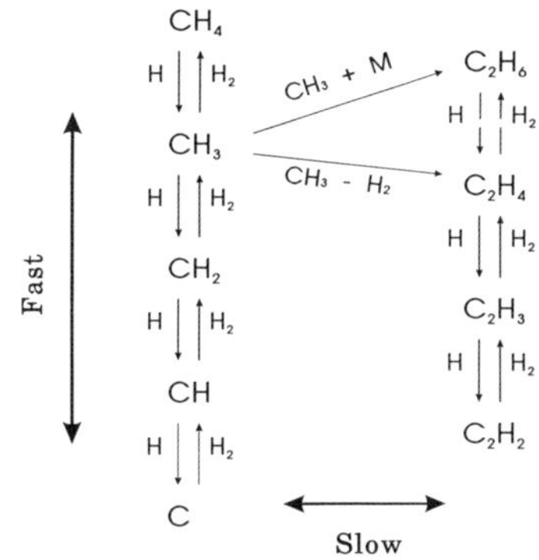

Figure 2. The principal gas phase reactions involve the rapid hydrogen transfer reactions amongst the C_1 and C_2 species, and to a lesser degree, the bimolecular hydrocarbon reactions forming C_2 and higher species.

chemistry is driven by reactions of atomic hydrogen with hydrocarbon species and reactions among the hydrocarbon species. A schematic of the major elements of this complex mix of reactions is depicted in figure 2, where reactions of larger than C_2 hydrocarbons are omitted. It is important to note that at pressures below 1 atm hydrogen atom recombination is slow, and a super-equilibrium concentration of hydrogen atoms is present (Goodwin 1991). Since the major species are H and H_2, and the total hydrocarbon concentration is less than 1%, the hydrogen transfer reaction rates are generally much greater than those describing the bimolecular hydrocarbon reaction rates (Harris 1989; Goodwin & Gavillet 1990; Franklach & Wang 1991; Coltrin & Dandy 1993).

The hot filament or plasma that is used to activate the deposition process is generally remote from the deposition surface. Transport from the activation or mixing zone to the near surface region is by diffusion and/or convection. The transport mechanism and geometry are often unique to a specific deposition technology, but common to most methods is a transport time much longer than the characteristic time for the hydrocarbons to react with hydrogen atoms. As a result those reactants coupled by hydrogen transfer reactions rapidly equilibrate and their distribution depends directly on the atomic to molecular hydrogen ratio. Subsequent reactions among hydrocarbon species may be near or far from equilibrium depending upon pressure and residence time since hydrocarbon concentrations are quite low.

Within several gas phase reactive mean free paths of the surface, the composition of the gas is perturbed by the effect of reactions occurring at the surface, forming a chemical boundary layer through which species diffuse. When the gas phase is dominated by convectively driven transport, as in plasma jets or torches, a fluid or momentum boundary layer is also formed. In these situations, only a small fraction of the gaseous species generated actually reach the deposition surface. The chemical composition of the flux of species arriving at the deposition surface may differ from the bulk gas well away from the surface due to chemical reactions caused by the temperature and concentration gradients in the boundary layer.

This simplistic overview of the deposition process is consistent with *in situ* laser probing of the gaseous species between the hot filament and the chemical boundary layer (Celii & Butler 1988, 1989; Chen *et al.* 1992) and recent fluid transport and

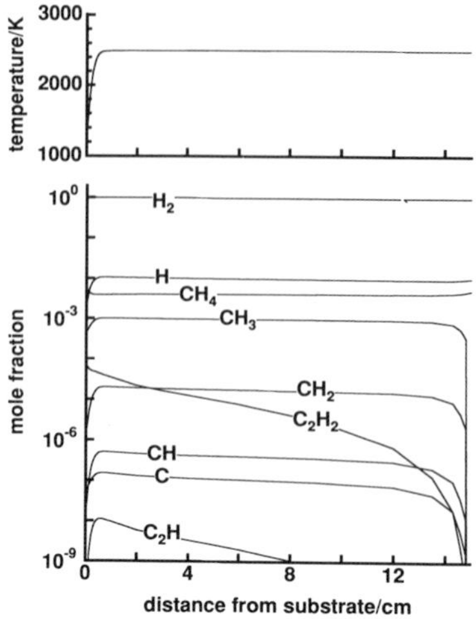

Figure 3. The gas phase species mole fraction profiles and temperature calculated for injection of CH_4 (0.5 %) into a H_2 flow containing 1 % atomic hydrogen. Initial flow velocity is 2×10^5 cm s^{-1}, pressure is 30 Torr and temperature is 2500 K. CH_4 injection is 15 cm upstream from the deposition surface.

chemical kinetic modelling of the growth environment which include surface chemistry (Coltrin & Dandy 1993). Results of such a model calculation are presented in figure 3 (1 % H results) and illustrate the difference in the gas composition in the near surface region versus the bulk gas phase (Dandy & Coltrin 1993). These results also demonstrate the tightly coupled kinetics of the single carbon species, atomic and molecular hydrogen, all of whose mole fractions are essentially constant throughout the transport process until they reach the near surface region. The two carbon species, however, are constantly increasing up to the near surface region.

3. Boundary layer transport

The presence of a chemical boundary layer above the surface has a profound effect on the arrival rate of reactants to the surface, and hence can severely limit growth rates. Consider the case of hydrogen atoms diffusing through a gas consisting mainly of molecular hydrogen at 30 Torr. If the surface reacts with all the incident atomic H, the gas phase concentration in the near surface region approaches zero and a large H atom gradient is developed with respect to the distance from the surface, thus defining a diffusion boundary layer (Kays & Crawford 1980). A lower limit on the H atom flux to the surface can be estimated from the diffusion constant D ($\approx$ 1100 cm^2 s^{-1}) and the diffusion distance d ($\approx$ 6 mm) in a hot filament or microwave plasma reactor. An approximate value for the diffusion time is $t\,(= d^2/2D$ or 0.16 ms in this case), which results in an average diffusion velocity of $v = d/t$ or 3750 cm s^{-1}, and a flux to the surface $J = n_{\mathrm{H}} v = 9 \times 10^{18}$ atoms cm^{-2} s^{-1} (n_{H} = H atom density). For the case where the surface is unreactive to the H atoms, the upper limit on the H atom flux can be computed from gas kinetics, $J = \frac{1}{4} n_{\mathrm{H}} c = 3 \times 10^{20}$ atoms cm^{-2} s^{-1} (where $c = 5 \times 10^5$ cm s^{-1} is the thermal velocity). Thus the presence of the chemical boundary layer can reduce the H atom flux by a factor of 30 and alter the concentrations of other chemical species which are reactively coupled to the atomic H.

Similarly, a forced convective flow, as is encountered in plasma jets and torches, can form a momentum boundary layer through which the reactive species must diffuse. The thickness of this momentum boundary layer is dependent on the fluid mechanics and viscosity (composition and pressure). High rate deposition processes, such as DC arc-jets, differ in both the fraction of dissociated hydrogen and the diffusion distance. Arc jets can have dissociation fractions over a factor of 10 higher than hot filament or microwave systems (typically 0.001 to 0.01 for a microwave system (Hsu 1991*a*, *b*)) and the high directed velocity of an arc jet can thin the boundary layer by a factor of 10, thereby increasing the hydrogen atom flux to the surface by over a factor of 100.

4. Surface processes

(*a*) *Surface structure*

The growth of the diamond lattice is a chemical process which occurs at the diamond surface. This surface is thought to be hydrogen terminated, like an alkane, since this is the usual condition for most natural diamonds and because of the abundance of hydrogen atoms in the CVD environment (Celii & Butler 1991). For bulk material, the graphite phase of carbon is slightly more stable than the diamond phase, but the interconversion is prevented by a large activation barrier. When one considers the local environment of the surface or interface where growth occurs, the surface stabilization by hydrogen can bias the local thermodynamics in favour of diamond growth (Yarbrough 1991). Thus central to the models of diamond CVD are the surface reactions with gaseous atomic hydrogen, hydrocarbon radicals and molecules, and molecular hydrogen.

Most of the surface hydrogen on diamond can be desorbed when the surface is subjected to temperatures in excess of 1100 K in ultra-high vacuum. Rearrangement of the surface atoms to form a lower symmetry reconstruction (2×1) has been observed for the nude (111) and (100) surfaces (see Celii & Butler 1991). Small amounts of non-hydrogen species (F, Cl, O, S, N) have also been detected on the diamond surface, usually after acidic treatments (see Celii & Butler 1991). Complete coverage of the diamond surface by non-hydrogenic species is usually precluded by steric effects due to the short bond length in the diamond lattice and resultant high density of surface atoms. Direct determination of the diamond surface in CVD environments is currently under study.

We present two surface chemical growth models. The first model, a generic model, addresses the basic chemical processes that lead to diamond growth and defect incorporation, without distracting the discussion with the important distinctions relating to the surface stereochemistry and the specific nature of the reacting gaseous hydrocarbon species. The second model, growth at the 110 site using $CH_3^{\bullet}$ radicals, incorporates the surface stereochemistry and focuses on a local surface site structure, which homoepitaxial growth studies and observed polycrystalline surface morphologies suggest is the predominant growth site. Growth from predominantly gaseous CH_x $(x < 4)$ species is strongly suggested from various experimental studies, and we use the methyl radical, $CH_3^{\bullet}$, as the example for the model.

Table 1. *Rate coefficients used in the site fraction and quality calculations*
(The units of A are given in terms of moles, cubic centimetres and seconds. The rate coefficient is expressed as $AT^n \exp(-E_a/RT)$.)

rate coefficient	A	n	$E_a/(\text{kcal mol}^{-1})$	reference
k_1	7×10^{13}	0	7.3	Westbrook *et al.* (1988)
k_2	1×10^{14}	0	0	Harris (1990)
k_3	1×10^{13}	0	0	Harris (1990)
k_{14}	1.12×10^{16}	-1	48	

(*b*) *Generic surface model*

The importance of atomic hydrogen for CVD diamond growth is seen by considering the deposition reactions of figure 2 in more detail:

$$C_DH + H^{\bullet} \rightleftharpoons C_D^{\bullet} + H_2, \tag{1}$$

$$C_D^{\bullet} + H^{\bullet} \rightarrow C_DH, \tag{2}$$

$$C_D^{\bullet} + C_xH_y \rightleftharpoons C_D\text{–}C_xH_y. \tag{3}$$

Reaction (1) represents activation of a surface site by removal of a surface hydrogen atom from a carbon atom on the diamond surface (C_D). An activated surface site is required for hydrocarbon addition. Either a radical (e.g. $CH_3^{\bullet}$) or unsaturated molecule (e.g. C_2H_2) may add to a radical site. Under conditions of diamond CVD, reactions (1)–(3) are fast and the reactions are in steady-state. Assuming that the H abstraction rate $k_1[C_DH][H^{\bullet}]$ is much greater than the unimolecular desorption rate $k_{-3}[C_D\text{–}C_xH_y]$, and applying steady-state to $[C_D^{\bullet}]$ results in equation (4) for the fraction of open sites

$$\frac{[C_D^{\bullet}]}{[C_DH]} = \frac{k_1[H^{\bullet}]}{k_{-1}[H_2]+k_2[H^{\bullet}]+k_3[C_xH_y]}. \tag{4}$$

Values for the rate coefficients in equation (4) are given in table 1. Examination of the terms in equation (4) for conditions typical of diamond CVD show that k_{-1} is small (k_{-1} is calculated from k_1 and the equilibrium constant for reaction (1)) so $k_{-1}[H_2]$ is much less than $k_2[H^{\bullet}]$. The concentration of reactive hydrocarbons $[C_xH_y]$ is usually less than $[H^{\bullet}]$ so $k_3[C_xH_y]$ is less than $k_2[H^{\bullet}]$. Thus equation (4) can be rewritten in terms of k_1 and k_2 only,

$$\frac{[C_D^{\bullet}]}{[C_DH]} = \frac{k_1}{k_2}. \tag{5}$$

In the limit of small hydrocarbon and high atomic hydrogen mole fractions the fraction of open sites depends only on temperature. The fraction of open sites as a function of temperature is shown in figure 4.

Clearly, the dominant surface process occurring is the conversion of two hydrogen atoms to one hydrogen molecule through reaction (1) and (2). Surface heating via surface mediated recombination of atomic hydrogen can be a significant contribution to heat flow in the diamond CVD process. This mechanism has been corroborated by recent photoionization mass spectrometric measurements of the rate of loss of atomic hydrogen to the diamond surface, which measure an activation barrier of

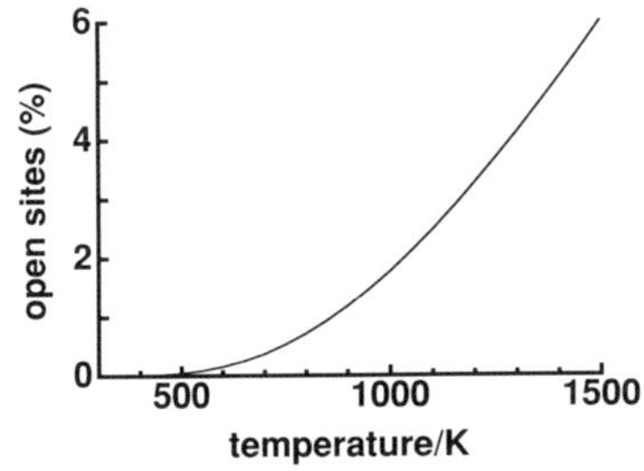

Figure 4. The fraction of open (hydrogen vacancy) sites on the hydrogenated surface against temperature is computed using equation (4) with the values in table 1.

6.7 kcal mol^{-1}† for rate coefficient k_1, consistent with hydrogen abstraction from hydrocarbons (Gutman, personal communication).

The rate of adding carbon to the surface, reaction (3), is thus independent of atomic hydrogen concentration. The reverse of the carbon addition, k_{-3}, is the unimolecular desorption of the hydrocarbon species and is important at higher temperature. However, when the hydrocarbon species has more than a single bond to the lattice carbons, the unimolecular desorption process is negligible.

Upon addition of a hydrocarbon radical or molecule to the surface, the newly added carbon has one bond to the lattice. To be fully incorporated, at least two more bonds must be made (assuming the fourth bond will be to the next layer to be deposited). Since the hydrocarbons tend to be hydrogenated, incorporation must involve removal of hydrogen. In most CVD diamond processes atomic hydrogen is the primary reagent for this process, analogous to reaction (1). Hence, we shall define *growth* of the lattice as the formation of at least two bonds between the hydrocarbon adsorbates and the diamond lattice, as opposed to the process we have considered thus far, namely the *addition* of a carbon species to the surface.

Now let us consider the kinetics of this chemisorbed hydrocarbon species and an adjacent hydrogen terminated surface site, $CH\cdots HC_D$

$$CH\cdots HC_D + H^{\bullet} \rightleftharpoons C^{\bullet}\cdots HC_D + H_2, \tag{6}$$

$$C^{\bullet}\cdots HC_D + H^{\bullet} \rightarrow CH\cdots HC_D, \tag{7}$$

$$C^{\bullet}\cdots HC_D + H^{\bullet} \rightarrow C^{\bullet}\cdots C^{\bullet}_D, \tag{8}$$

$$C^{\bullet}\cdots C^{\bullet}_D \rightarrow C\text{–}C_D. \tag{9}$$

Using similar assumptions as above and noting the reaction (6) and (8) are the same as reaction (1), and reaction (7) is the same as reaction (2), we consider steady state on the $C^{\bullet}\cdots HC_D$ species to obtain equation (10), which is independent of atomic hydrogen concentration analogous to equation (5).

$$\frac{[C^{\bullet}\cdots HC_D]}{[CH\cdots HC_D]} = \frac{k_1}{k_1 + k_2}. \tag{10}$$

Incorporation of carbon into the diamond lattice is represented by the sum of equations (8) and (9), and shows that the rate of lattice incorporation depends directly on the atomic hydrogen concentration.

$$C^{\bullet}\cdots HC_D + H^{\bullet} \Rightarrow C\text{–}C_D + H_2. \tag{11}$$

† 1 cal = 4.184 J.

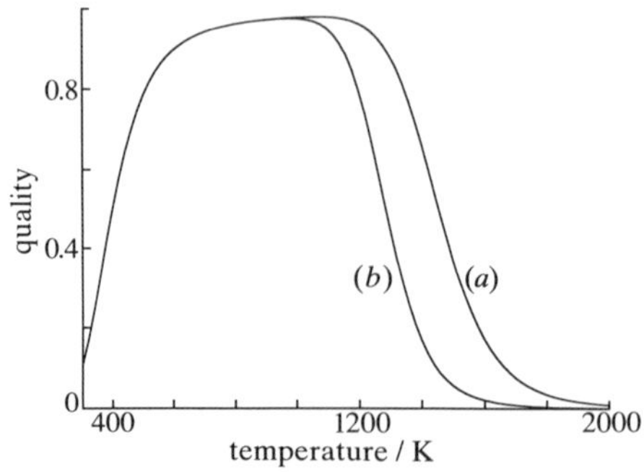

Figure 5. The calculated quality (lattice incorporation/carbon addition) is shown as a function of temperature for conditions characteristic of (*a*) arc jet ($X_H = 0.1$, $X_{CxHy} = 10^{-4}$) and (*b*) hot filament ($X_H = 0.01$, $X_{CxHy} = 10^{-5}$) reactors.

(*c*) *Defects*

Growth as we have defined it includes both the growth of the sp^3 diamond lattice as well as the potential for the incorporation of sp^2 defects formed by unimolecular decomposition of $C^{\cdot}\cdots HC_D$, as well as $CH\cdots HC_D$ and related species. In this process the growth or hydrogen abstraction steps were not completed on the timescale in which competing adjacent sites were able to overgrow the specific surface species in question. If additional layers of carbon deposit prior to complete C–C bonding, then the missing lattice bond will be buried and represent a defect. The sp^3 defects will be characterized by hydrogen terminated sp^3 carbon, with a characteristic CH stretching infrared absorption spectrum (2800–3000 cm^{-1}) (Willingham *et al.* 1991). (Note that a completely missing carbon atom in the lattice will give rise to a similar hydrogen terminated defect.) Unimolecular decomposition of the $C^{\cdot}\cdots HC_D$ moiety (which may desorb a hydrocarbon) to create sp^2 lattice bonds will form a second type of defect with the characteristic sp^2 Raman spectrum observed at elevated temperatures.

Equation (11) represents incorporation of the desired sp^3 bonds in the lattice. In competition with sp^3 incorporation are further addition (equation (12)), desorption (equation (13)), and sp^3 defect formation (equation (14)).

$$C^{\cdot}\cdots HC_D + C_xH_y \Rightarrow C_xH_y\text{–}C\cdots HC_D, \tag{12}$$

$$C^{\cdot}\cdots HC_D \Rightarrow C_DH + C_2H_y, \tag{13}$$

$$C^{\cdot}\cdots HC_D \Rightarrow C_{defect}\cdots HC_D. \tag{14}$$

Defining quality as the fraction of lattice bonds which are sp^3, we arrive at equations (15) and (16), recognizing that $k_{12} = k_3$

$$\text{quality} = \frac{[C\text{–}C_D]}{[C\text{–}C_D] + [C_xH_y\text{–}C\cdots HC_D] + [C_{defect}]}, \tag{15}$$

$$\text{quality} = \frac{k_1[H^{\cdot}]}{k_1[H^{\cdot}] + k_3[C_xH_y] + k_{14}}. \tag{16}$$

Calculation of this quality factor for several atomic hydrogen concentrations and a C_xH_y mole fraction of 10^{-4} is shown in figure 5. This generic model demonstrates that under conditions typical for most diamond CVD (i) the fraction of open sites on

Figure 6

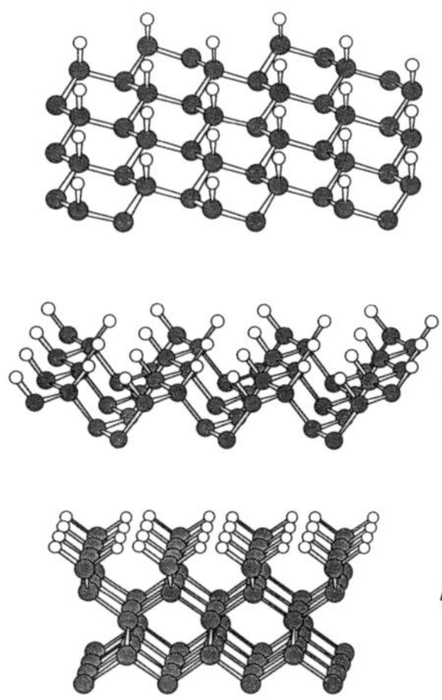

Figure 7

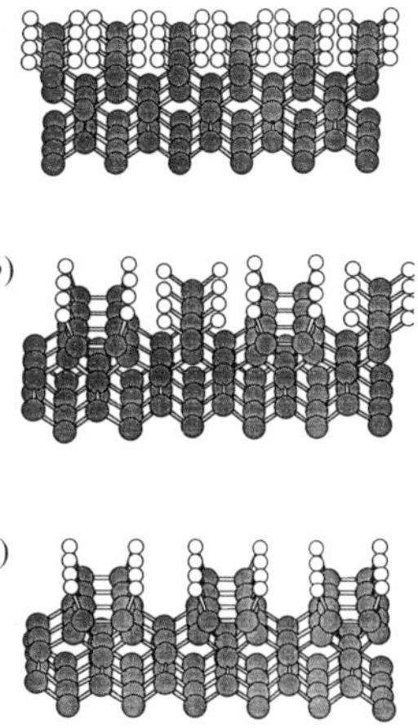

Figure 6. The hydrogen truncation of the (111), (110), and (100) surfaces of the diamond lattice is shown. The dark large circles represent the near surface carbon atoms, and the smaller open circles, the terminating hydrogen atoms.

Figure 7. Three ways (of many possible) of hydrogen truncating the (100) surface. (*a*) Dihydride, (*b*) 50:50 dihydride: monohydride and (*c*) monohydride. (*b*) and (*c*) relax the severe steric constraints of (*a*).

the surface is independent of [H$^\cdot$] and dependent on temperature, (ii) the growth or incorporation of carbon into the lattice depends on [H$^\cdot$], (iii) the quality has a maximum and falls off at higher temperatures due to competing desorption and decomposition reactions, and (iv) the quality of the deposited diamond increases with [H$^\cdot$] and decreases with [C_xH_y]. These rules are generally consistent with a wide variety of observations made in diverse growth environments.

(*d*) *Surface morphology*

Surface structures of the hydrogen terminated diamond lattice are shown in figure 6 for the three low index planes, (111), (110), and (100). Nucleation (or growth) of the next layer requires the incorporation of three carbon atoms on the (111) surface, two on the (110), and only one on the (100). The structure shown in figure 6 for the (100) surface is problematic because adjacent hydrogen atoms are closer than they would be in molecular hydrogen, so clearly the structure must alter in some way to satisfy the steric constraints. While there has been no experimental determination of the structure of this (100) surface to date, figure 7 shows several additional structures which may be more stable. The full dihydride surface, figure 7*a*, is unlikely due to steric constraints, while the monohydride surface, figure 7*c*, distorts to form five member carbon rings between the surface and sub-surface carbons through the loss of pairs of hydrogen atoms. A more realistic structure may lie somewhere between these two extremes (Yang & D'Evelyn 1992) and is represented by the 50:50 dihydride:monohydride structure (figure 7*b*).

The diamond crystallites grown by CVD processes display (111) and (100) surface morphologies at low resolution, and at higher resolution (e.g. SEM, STM and AFM images) are seen to be rough, with many steps (Welland *et al.* 1992; Maguire *et al.*

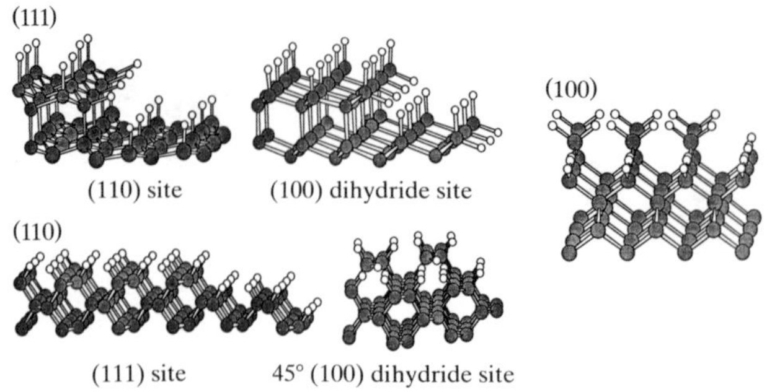

Figure 8. The local geometry (structure) of steps on the low index planes is shown.

1992). Therefore much of the growth surface is comprised of steps as well as crystalline planes. The local morphology or symmetry of steps on the low index planes is shown in figure 8. It is important to note that the local structure (or chemical environment) on a rough surface can be generated from a relatively small number of structures similar to the local structure of the low index planes. Multiple twin planes parallel to the (111) plane are common imperfections observed in CVD diamonds (Williams *et al.* 1990). The intersections of these twin planes with the surfaces can cause re-entrant sites similar to the (110) trough structure and have been proposed as likely growth sites (Angus *et al.* 1989).

Let us consider what is happening to the CH surface terminating species in the CVD environment. The typical surface coverage of hydrogen on the diamond surface, S, is *ca.* 1×10^{15} sites cm^{-2}. Using the atomic hydrogen fluxes derived earlier, this leads to an upper limit on the collision frequency, f, of gaseous H with a surface H site of $f = J/S \leqslant 3\times10^{5}\ s^{-1}$. Choosing an abstraction probability of 0.03 to 0.1 at the 1200 K gas and surface temperature, suggests that each surface H is abstracted to create a vacancy on a time scale greater than 60–70 μs. The usual fate of such a vacancy will be reaction with another incoming H atom. The open time of a surface hydrogen vacancy (radical site) is the reciprocal of the collision frequency (f) divided by the probability of attachment (*ca.* 0.4) (Brenner *et al.* 1993), which will be greater than 8 μs. At the elevated temperatures typical of CVD (1000–1300 K), surface hydrogen vacancies are likely mobile in certain directions, e.g. along the (110) troughs by hopping of the hydrogen from the opposite side, with hopping frequencies estimated around one per microsecond (D. G. Goodwin, personal communication). Hydrogen vacancy mobility has also been proposed along the (100) di-hydride direction (Huang & Frenklach 1992).

It is apparent that the surface of diamond in the growth environment is very active with the abstraction of each surface hydrogen generating a vacant site every 70 μs (on average), and the refilling of the site 15 μs later, allowing for the potential of surface hydrogen hopping or vacancy diffusion along certain directions, and relatively short mean distances between adjacent vacancies.

(*e*) CH_3 *growth at* 110 *sites*

A number of models have been developed for the growth steps evolving from the addition of a hydrocarbon species to the surface to the incorporation of the carbon into the lattice (see Tsuda *et al.* 1986; Frenklach & Spear 1988; Harris 1990; Belton

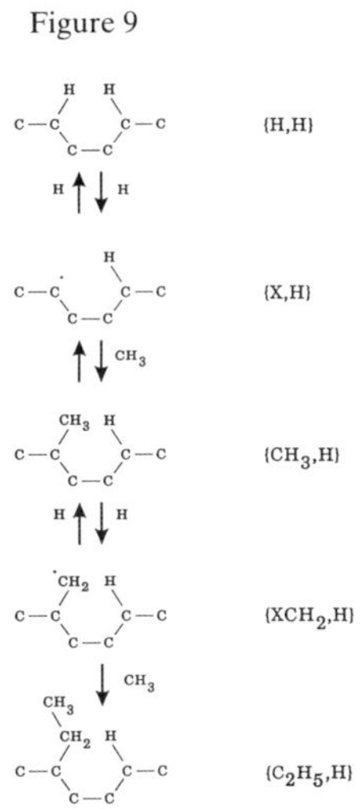

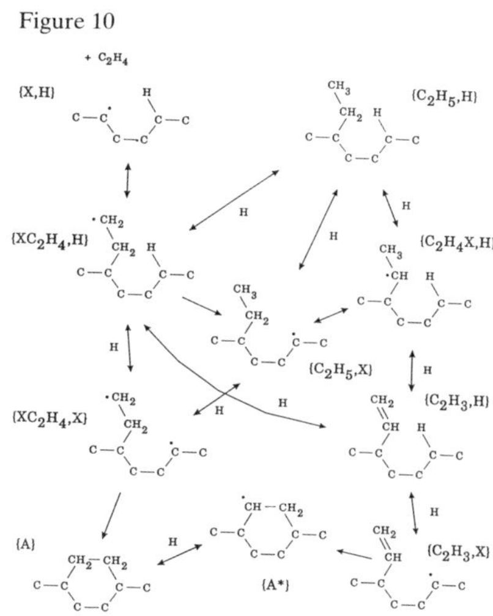

Figure 9. The reaction path for hydrogen abstraction, retermination, and eventual growth of a hydrocarbon species singly bonded to the surface is shown for a representation of the (110) trough site starting with two adjacent hydrogen species, {H, H}. The (110) troughs are shown in perspective in figure 6.

Figure 10. Several possible fates for the adsorbed ethyl species {C_2H_5, H} are shown leading to the desorption of C_2H_4, an sp^3 bonded CH_2–CH_2 bridge species {A}, or an ethylene group singly bonded to the surface, {C_2H_3, H}.

& Harris 1992; Garrison *et al.* 1992). In this discussion, we shall not attempt to cover all the proposed alternatives, but focus on a particular pathway which may account for most of the carbon incorporation into the lattice: a mechanism for growth by single carbon species (e.g. CH_3) at (110) type growth sites. The emphasis on the (110) site is based on the (110) surfaces showing the high growth rate in homo-epitaxial growth studies (Chu *et al.* 1991, 1992), the (110) growth texture observed in many polycrystalline films (the fastest growth direction) (Wild *et al.* 1990; Specht *et al.* 1991), the lack of macroscopic (110) faces on the crystallites due to those facets growing out (cubo-octahedral morphology is usually observed), and the (110) local structure exhibited by steps and re-entrant edges, where twin planes intersect the surface as discussed above. The CH_3 radical is emphasized as the growth species because of the kinetic and spectroscopic data indicating that it (or at least a single carbon species) is responsible for most of the carbon incorporated into the diamond lattice.

The first step of this mechanism, as in the generic model, is the generation of a chemisorbed hydrocarbon (ethyl) group on the surface through successive attachments of CH_x ($x < 4$) species (methyl radicals) at radical sites, as shown in the cartoons of the (110) trough site in figure 9. Note that most of the steps in the mechanism proposed here involve the facile hydrogen abstraction and re-attachment, the occasional chemisorption of the hydrocarbon species at a radical site, unimolecular desorption of hydrocarbon species singly bonded to the surface at higher temperatures, and facile radical–radical reactions on the surface. Mobility of surface hydrogen vacancies or radical sites can also assist many of the steps

Figure 11

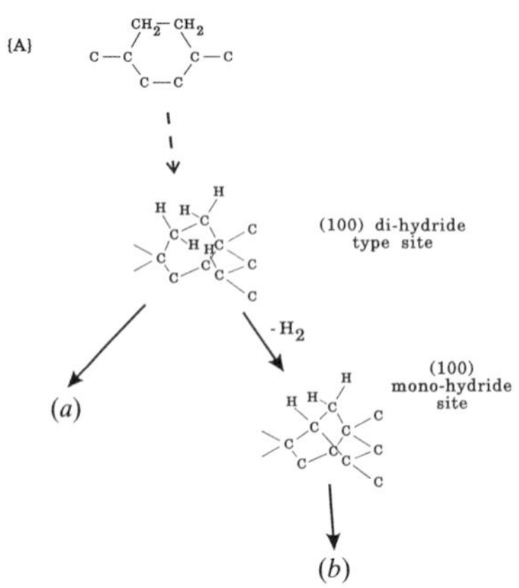

Figure 12

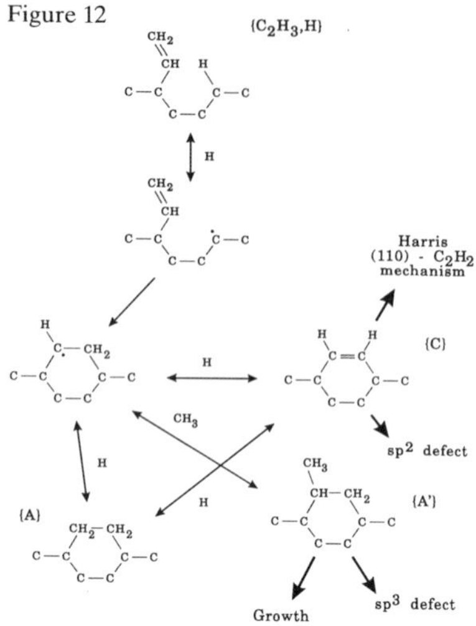

Figure 11. The CH_2–CH_2 bridge species {A} provides either a (100) dihydride site or desorbs hydrogen to form a (100) monohydride site for continued growth from C_1 species along the direction of the (110) trough. Mechanisms for such growth have been proposed for the (100) surface by Harris (1990) or by Garrison *et al.* (1992) for the (100) monohydride site. (*a*) Harris (100) CH_3 mechanism. (*b*) Garrison *et al.* (100) monohydride site and CH_3 mechanism.

Figure 12. The ethylene group singly bonded to the surface $\{C_2H_3, H\}$ can desorb C_2H_2, convert to the bridged species {A}, form an sp^2 bonded bridge {C}, or a modified sp^3 bonded bridge species {A′}. The sp^2 bonded bridge species may be overgrown by adjacent sites to form a defect, or grow into the lattice when there are adjacent sp^2 bridges by the Belton and Harris (110) C_2H_2 mechanism (Belton & Harris 1992). The modified sp^3 bonded bridge species {A′} can be overgrown to form an sp^3 hydrogen defect or continue growth of the lattice.

discussed. The fate of the chemisorbed ethyl is shown in figure 10. One pathway leads to the desorption of ethylene via a beta-scission reaction after the abstraction of a hydrogen atom, $\{C_2H_5, H\} \rightarrow \{XC_2H_4, H\} \rightarrow \{X, H\} + C_2H_4$ in figure 10, thus providing a surface mechanism for the conversion of gaseous C_1 species to C_2 species. Many other combinations of processes lead to the C_2H_4 bridged species, {A} in figure 10, and a few lead to the chemisorbed vinyl group, $\{C_2H_3, H\}$ in figure 10. The fates of the chemisorbed vinyl group are discussed below, and one of these, conversion to the bridged species {A} is shown in figure 10.

A look at the conformation of C_2H_4 bridged species, {A}, shows that (100) dihydride type sites exist at each edge of the bridge, as is shown in figure 11. This bridge structure is not as sterically constrained as the (100) surface and may thus survive for continued growth down the (110) troughs by the addition on single carbon species, e.g. methyl radicals. One mechanism already proposed by Harris (1990) adds methyl radicals to this (100) dihydride site. If the dihydride site is sufficiently unstable to the elimination of two hydrogens to form the (100) monohydride site, then another mechanism proposed by Garrison *et al.* (1992) can continue the growth along the trough by methyl radical addition and hydrogen abstraction.

The chemisorbed vinyl group, $\{C_2H_3, H\}$, is shown in figure 12. One reactive pathway for this species leads to the desorption of acetylene, while another leads to the C_2H_4 bridged species {A}. Two other outcomes are possible. The first generates a sp^2 bonded bridged species, {C} in figure 12. If the {C} species are generated adjacent to one another, which seems unlikely in this environment, they can zip up to form the next layer of the diamond lattice, as proposed in the Belton and Harris mechanism for C_2H_2 growth at (110) sites (Belton & Harris 1992). Otherwise this species can be hydrogenated to form {A}, or overgrown by adjacent sites leading to a sp^2 defect in the lattice. The second outcome is a modified sp^3 bridged species which can continue growth of the lattice or also be overgrown to provide an sp^3 defect.

While most of the rate constants for these diamond surface reactions are unknown, either experimentally or from theoretical computation, the types of reactions proposed are well known in organic and combustion chemistry, and reasonable estimates of key rate constants can be made to estimate the order of magnitude of the total growth rate. This has not been done yet. What is clear is that the generation of the bridged species {A} requires the incorporation of at least two low probability events, e.g. the methyl radical reaction at a surface radical site. However, once this bridged site is generated, further growth can occur by the addition of single carbon events. In other words, *an infrequent nucleation of a growth site is followed by more frequent growth events around that site.*

5. Discussion

In this paper we have attempted to outline the principal steps transforming gaseous hydrocarbons into diamond crystals under conditions of temperature and pressure in which bulk diamond is metastable relative to graphite. There are many nuances and variations to this process which could not be explicitly addressed without greatly expanding (and distracting) this discussion. A principal variation that is commonly used is the addition of oxygen containing species to the reactants. This leads to Bachmann *et al.*'s (1991) C–H–O phase diagram for diamond CVD. This phase diagram can be rationalized chemically by recognizing that the rapid formation of CO from the reactants in the plasma, flame, of hot filament environment results from extreme chemical stability of the CO molecule. Once formed, the CO will be inert unless it is still in the activation plasma or flame front. Hence, when the transport time from the activation zone to the growth surface is long enough, the real flux of growth species to the surface can be approximated by comparison to a C–H environment where the [C]/[H] ratio is ([C]−[O])/[H], in which [C], [H], and [O] represent the sums of all the atoms comprising the reactants. For example, a filament assisted CVD process using 1% O_2, 2.5% CH_4 (or 1.25% C_2H_4) in H_2 would, at the first level of approximation, be similar to a process using 0.5% CH_4 (or 0.25% C_2H_4) in H_2.

The principal focus of this presentation is the role of the gaseous and surface chemistry of H atoms in diamond synthesis. However, other species can function as surrogates for the H atom in each of the proposed roles. Likely species are OH, F, Cl, and O, and indeed, variations of diamond CVD growth environments exist in which these can exist as species (though often in minor concentrations). Since the reactions of these species have different thermodynamics and kinetics, and particularly, different activation energies, significant effects are observed on properties which are controlled by a delicate balance of several processes, e.g. crystal morphology, defect concentrations and types, and the temperature limits to growth.

In the generic surface chemistry model outlined above, atomic hydrogen does not enter explicitly into the addition rate of carbon to the surface. It does however, control the rate of growth of the chemisorbed hydrocarbon species into the lattice (or dehydrogenation) and is necessary to 'heal' defects (either by C–C bond formation or hydrogenation of sp^2 bonding). Thus the quality of the diamond should depend directly on atomic hydrogen flux to the surface. Conversely, depositing carbon faster than it can be grown into the lattice, or defects can be healed, will lead to defect incorporation into the bulk of the material and the quality will be degraded. As a result, for a given atomic hydrogen concentration higher quality diamond will be deposited at lower growth rates. To increase growth rate at high quality, higher atomic hydrogen concentrations will be required.

6. Summary

This paper has highlighted the principal processes occurring during the chemical vapour deposition of diamond. The reactants are activated by a hot-filament or plasma, whose principal role is the production of atomic hydrogen. The atomic hydrogen drives subsequent gaseous reactions amongst the dilute hydrocarbons while these species are transported to the deposition surface by diffusion and convection. At the surface, the atomic hydrogen terminates and stabilizes the carbon lattice by forming CH bonds, generates radical sites at which hydrocarbons can adsorb by hydrogen abstraction, and assists the incorporation of the hydrocarbon species into the lattice by further hydrogen abstraction. Two models of the surface chemistry are presented. The first, a generic model ignoring stereochemistry, demonstrates that (i) the fraction of surface radical sites is a function of temperature only and does not depend on the atomic hydrogen flux, (ii) the growth of adsorbed hydrocarbons into the lattice does depend on the atomic hydrogen flux, (iii) diamond growth is limited by hydrogen abstraction at low temperatures and by thermal desorption of adsorbates at higher temperatures, and (iv) the quality is proportional to the hydrogen flux and inversely proportional to the reactive hydrocarbon flux. The second model addresses the local chemical environment (geometry) of the growth site on the surface and proposes a growth mechanism at (110) type of sites from C_1 hydrocarbon species, e.g. methyl radicals. The potential for the incorporation of sp^2 and sp^3 defects in the deposited material is addressed in both models.

This work was supported in part by the Office of Naval Research (ONR) and the Defense Advanced Research Projects Agency (DARPA). Helpful and illuminating discussions with D. Dandy, M. Coltrin, D. Goodwin, J. N. Russell Jr, V. Shamamian and M. McGonigal are greatly appreciated. We thank E. Bier for assistance with the figures.

References

Angus, J. C., Buck, F. A., Sunkara, M., Groth, T. F., Hayman, C. C. & Gat, R. 1989 Diamond growth at low pressures. *Mater. Res. Soc. Bull.* October, 33–47.

Bachmann, P. K., Leers, D. & Lydtin, H. 1991 Towards a general concept of diamond chemical vapor deposition. *Diamond Related Mater.* **1**, 1–12.

Belton, D. N. & Harris, S. J. 1992 A mechanism for growth on diamond (110) from acetylene. *J. chem. Phys.* **96**, 2371–2377.

Brenner, D. W., Robertson, D. H., Carty, R. J., Srivastava, D. & Garrison, B. J. 1993 Combining molecular dynamics and monte carlo simulations to model chemical vapor deposition: applications to diamond. *Proc. Mater. Res. Soc.* (In the press.)

Celii, F. G. & Butler, J. E. 1989 Hydrogen atom detection in the filament-assisted diamond deposition environment. *Appl. Phys. Lett.* **54**, 1031–1033.

Celii, F. G. & Butler, J. E. 1991 Diamond chemical vapour deposition. *A. Rev. phys. Chem.* **42**, 643–684.

Celii, F. G. & Butler, J. E. 1992 Direct monitoring of CH_3 in a filament-assisted diamond chemical vapor deposition reactor. *J. appl. Phys.* **71**, 2877–2883.

Celii, F. G., Pehrsson, P. E., Wang, H. T. & Butler, J. E. 1988 Infrared detection of gaseous species during the filament-assisted growth of diamond. *Appl. Phys. Lett.* **52**, 2042–2045.

Chen, K.-H., Chuang, M.-C., Penney, C. M. & Banholzer, W. F. 1992 Temperature and concentration distribution of H_2 and H atoms in hot-filament chemical vapor deposition of diamond. *J. appl. Phys.* **71**, 1485–1493.

Chu, C. J., D'Evelyn, M. P., Hauge, R. H. & Margrave, J. L. 1991 Mechanism of diamond growth by chemical vapor deposition on diamond (100), (111), and (110) surfaces: carbon-13 studies. *J. appl. Phys.* **70**, 1695–1705.

Chu, C. J., Hauge, R. H., Margrave, J. L. & D'Evelyn, M. P. 1992 Growth kinetics of (100), (110), and (111) homoepitaxial diamond films. *Appl. Phys. Lett.* **61**, 1393–1395.

Coltrin, M. E. & Dandy, D. S. 1993 An elementary reaction mechanism for the growth of diamond. *J. appl. Phys.* (In the press.)

Dandy, D. S. & Coltrin, M. E. 1993 Analysis of diamond growth in subatmospheric dc plasma-gun reactors. *J. appl. Phys.* (In the press.)

Frenklach, M. & Spear, K. E. 1988 Growth mechanism of vapor-deposited diamond. *J. Mater. Res.* **3**, 133–140.

Frenklach, M. & Wang, H. 1991 Detailed surface and gas-phase chemical kinetics of diamond deposition. *Phys. Rev.* B **43**, 1520–1545.

Garrison, B.J., Dawnkaski, E. J., Srivastava, D. & Brenner, D. W. 1992 Molecular dynamics simulations of dimer opening on a diamond {001} (2×1) surface. *Science, Wash.* **255**, 835–838.

Goodwin, D. G. 1991 Simulations of high-rate diamond synthesis: Methyl as growth species. *Appl. Phys. Lett.* **59**, 277–279.

Goodwin, D. G. & Gavillet, G. G. 1990 Numerical modeling of the filament-assisted diamond growth environment. *J. appl. Phys.* **68**, 6393–6400.

Harris, S. J. 1989 Gas-phase kinetics during diamond growth: CH_4 as-growth species. *J. appl. Phys.* **65**, 3044–3048.

Harris, S. J. 1990 Mechanism for diamond growth from methyl radicals. *Appl. Phys. Lett.* **56**, 2298–2300.

Harris, S. J., Weiner, A. M. & Perry, T. A. 1988 Filament-assisted diamond growth kinetics. *Appl. Phys. Lett.* **53**, 1605–1607.

Huang, D. & Frenklach, M. 1992 Energetics of surface reactions on (100) diamond plane. *J. phys. Chem.* **96**, 1868–1875.

Hsu, W. L. 1991*a* Quantitative analysis of the gaseous composition during filament-assisted diamond growth. *Electrochem. Soc. Proc.* **91**, 217–223.

Hsu, W. L. 1991*b* Mole fraction of H, CH_3, and other species during filament-assisted diamond growth. *Appl. Phys. Lett.* **59**, 1427–1429.

Kays, W. M. & Crawford, M. E. 1980 *Convective heat and mass transfer*, ch. 4, 2nd edn. New York: McGraw-Hill.

Maguire, H. G., Kamo, M., Lang, H. P., Meyer, E., Weissendanger, K. & Guntherodt, H. J. 1992 The structure of conducting and non-conducting homoepitaxial diamond. *Diamond Related Mater.* **1**, 634–638.

Mucha, J. A., Flamm, D. L. & Ibbotson, D. E. 1989 On the role of oxygen and hydrogen in diamond-forming discharges. *J. appl. Phys.* **65**, 3448–3452.

Specht, E. D., Clausing, R. E. & Heatherly, L. 1991 X-ray and optical characterization of three growth morphologies of CVD diamond films. *J. Crystal Growth* **114**, 38–46.

Tsuda, M., Nakajima, M. & Oikawa, S. 1986 Epitaxial growth mechanism of diamond crystal in CH_4–H_2 plasma. *J. Am. chem. Soc.* **108**, 5780–5783.

Welland, M. E., McKinnon, A. W., O'Shea, S. & Amaratunga, G. A. J. 1992 Scanning tunnelling microscopy and atomic force microscopy of carbon-diamond films. *Diamond Related Mater.* **1**, 529–534.

Westbrook, C. K., Warnatz, J. & Pitz, W. J. 1988 A detailed chemical reaction mechanism for the oxidation of isoctane and *n*-heptane over an extended temperature range and its application to analysis of engine knock. In *Proc. 22nd Symp. (Int.) on Combustion*, pp. 893–901. Seattle, Washington: The Combustion Institute.

Wild, C. H., Herres, N. & Koidl, P. 1990 Texture formation in polycrystalline diamond films. *J. appl. Phys.* **68**, 973–978.

Williams, B. E., Glass, J. T., Davis, R. F. & Kobashi, K. 1990 The analysis of defect structures and substrate/film interfaces of diamond thin films. *J. Crystal Growth* **99**, 1168–1176.

Willingham, C., Hartnett, T., Robinson, C. & Klein, C. 1991 Polycrystalline diamond for infrared optical applications prepared by the microwave plasma and hot filament chemical vapor deposition techniques. In *Applications of diamond films and related materials* (ed. Y. Tzeng, M. Yoshikawa, M. Murakawa & A. Feldman), pp. 157–162. Amsterdam: Elsevier.

Wu, C.-H., Tamor, M. A., Potter, T. J. & Kaiser, E. W. 1990 A study of gas chemistry during hot-filament vapor deposition of diamond films using methane/hydrogen and acetylene/hydrogen gas mixtures. *J. appl. Phys.* **68**, 4825–4829.

Yang, Y. L. & D'Evelyn, M. P. 1992 Structure and energetics of clean and hydrogenated diamond (100) surfaces by molecular mechanics. *J. Am. chem. Soc.* **114**, 2796–2801.

Yarbrough, W. A. 1991 Non-equilibrium thermodynamics and the vapour phase preparation of diamond for electronic applications. *J. electron. Mater.* **20**, 133–139.

Comment

L. M. Brown and P. Fallon (*Cavendish Laboratory, Cambridge, U.K.*). The picture presented is of a surface layer on the growing diamond consisting of hydrogen and a carbon in a variety of bonding states. It may therefore be of interest that using electron energy loss spectroscopy in the electron microscope we have observed thin (3 nm or less) layers of amorphous carbon at external surfaces and along grain boundaries in CVD diamond. The layers appear not to be hydrogenated, and we estimate them to be more than 90% sp^2 bonded. Although the external layers may be affected by the shut-down sequence of the plasma process, the intergranular layers are more likely to have arisen from the impingement of the layers on approaching surfaces, and to be kinetically stabilized because the interlayer is trapped and isolated from the plasma. It is probable that at the high growth temperatures the hydrogen initially contained in the layers evaporates.

3

Microwave plasma CVD and related techniques for low pressure diamond synthesis

BY PETER K. BACHMANN

The current status, perspectives and bottlenecks of the major diamond deposition techniques are compared. General trends and correlations are outlined. The emphasis is on plasma-assisted, low pressure routes of diamond synthesis.

---

1. Introduction

The preparation techniques known today for growing diamond from the vapour phase at low pressures and moderate temperatures range from thermally initiated chemical vapour deposition (CVD) methods to the application of a broad spectrum of plasma-induced CVD techniques. In table 1, these methods are categorized according to the specific way of initiating the chemical reactions and listed with the references of original articles (selected review articles are listed at the end of the reference list).

This review discusses the current status, advantages and drawbacks of most of the techniques for growing diamond from the vapour phase. They are compared method by method, but comparison of plasma-induced techniques is emphasized. The area of low pressure diamond CVD has experienced tremendous growth over the past 10 years and a review of the preparation technologies can only briefly summarize the present state of the art. However, the general trends and conclusions in the second part of this review together with the extensive reference list at its end will, hopefully, be a useful guideline for evaluating results and new developments in this field as well as providing a strategy tool to design new experimental approaches.

2. Low pressure diamond synthesis: techniques and results

(a) Thermal CVD methods and combustion flame synthesis

Early approaches to forming diamond from the vapour phase were characterized by the thermal decomposition of carbon-containing gases such as CBr_4, Cl_4 (Spitsyn & Derjaguin 1956), CO (Eversole 1962; Hibshman 1968), or CH_4 (Eversole 1962; Angus *et al.* 1968), carried out at gas temperatures between 600 and 1200 °C. The gas temperatures did not differ from the surface temperature of the diamond seeds that were exclusively used as substrates. The resulting linear diamond growth rates of approximately 0.01 $\mu m\ h^{-1}$ were far too low for any industrialization of diamond CVD. In addition, the material was contaminated with non-diamond carbon. After its initial boost in the 1950s and 1960s, the field lay dormant for more than a decade. However, during the revival of diamond CVD and with the advent of a wide variety of other CVD techniques, even such early approaches of what is now termed 'halogen-assisted' thermal CVD of diamond

Table 1. *Methods for synthesizing diamond at low pressures and low temperatures*

method	reference
thermal CVD	
thermal decomposition	Spitsyn & Derjaguin (1956); Eversole (1962); Hibshman (1968); Angus *et al.* (1968)
chemical transport reaction (CTR)	Spitsyn *et al.* (1981); Spitsyn (1991); Roy (1993); Bachmann (1993)
hot filament technique	Matsumoto *et al.* (1982); Sawabe & Inuzuka (1985); Hirose & Terasaki (1986); Sommer & Smith (1991)
oxy-acetylene torch	Hirose & Kondo (1988); Snail *et al.* (1991); Hirose *et al.* (1989); Matsui *et al.*(1991); Janssen *et al.* (1990); Doverspike & Butler (1993)
halogen assisted CVD	Spitsyn & Derjaguin (1956); Patterson *et al.* (1991)
DC plasma CVD	
low pressure DC plasma	Pinneo (1987); Ravi & Landstrass (1989); Thorsheim *et al.* (1991)
medium pressure DC plasma	Suzuki *et al.* (1987, 1990)
hollow cathode discharge	Singh *et al.* (1988)
DC arc plasmas and plasma jets	Ohtake *et al.* (1989); Matsumoto *et al.* (1988); Kurihara *et al.* (1988); Matsumoto (1988); Bachmann *et al.* (1990); Lu & Bigelow (1992)
radiofrequency plasma CVD	
low pressure RF glow discharge	Matsumoto (1985); Meyer *et al.* (1989); Wood *et al.* (1990); Rudder *et al.* (1991)
thermal RF plasma CVD	Matsumoto *et al.* (1987); Owano *et al.*(1991); Kohzaki *et al.* (1993)
microwave plasma CVD	
915 MHz plasma	Miyake *et al.* (1988)
2.45 GHz low pressure plasma	Kamo *et al.* (1983, 1990); Badzian *et al.* (1986); Bachmann *et al.* (1988*a*, *b*); Smith *et al.* (1992); Ishibori & Ohira (1990); Jiang & Klages (1993); Liou *et al.* (1988); Chang *et al.* (1988)
2.45 GHz thermal plasma torch	Mitsuda *et al.* (1989)
2.45 GHz magnetized (ECR) plasma	Kawarada *et al.* (1987); Suzuki (1989); Yuasa *et al.* (1991)
8.2 GHz plasma	Aklufi & Brock (1989)
other (non-CVD) methods	
C implantation with laser treatment	Prins & Gaigher (1991); Lau *et al.* (1992)
laser conversion of amorphous carbon	Fedoseev *et al.* (1983); Aslam *et al.* (1989)

(Spitsyn & Derjaguin 1956), i.e. to support diamond formation by adding halogen compounds to the CVD gas phase, were revisited. R. Hauge and co-workers (Patterson *et al.* 1991) demonstrated that 'halogen-assisted' thermal diamond CVD is capable of producing diamond at temperatures below 300 °C and that deposition on non-diamond substrates is feasible. To industrialize the simple thermal initiation of the diamond deposition reactions would be highly desirable: the design of the deposition system is straightforward and coating of three-dimensional bodies, a trouble spot for many plasma-induced methods, is fairly easy. This, in combination with the feature of growing at low deposition temperatures, would give halogen-assisted thermal CVD a clear advantage over many other methods of growing diamond. However, data on this technique are, even several years after its revival, still very limited. Reactor furniture corrosion in a hydrogen–halogen environment is certainly one of its major disadvantages. In addition, film formation (rather than particles) seems to be difficult with this method, the deposition rates are still low and the deposit is often heavily contaminated by metals that are etched away from the reactor walls.

Industrially applicable diamond CVD processes were within reach as soon as substrate surface temperatures and gas phase temperatures were decoupled (Spitsyn *et al.* 1981; Spitsyn 1991; Matsumoto *et al.* 1982; Kamo *et al.* 1983). The decomposition of the carbon carrier as well as the formation of the atomic hydrogen that is required as an agent to suppress or reduce the formation of non-diamond carbon is more effective at temperatures higher than the substrate temperature that still need to remain below ≈ 1300 °C to avoid graphitization of the growing diamond film. Both the ground-breaking 'chemical transport reaction' (CTR) method (Spitsyn *et al.* 1981; Spitsyn 1991) and the now widespread 'hot filament' technique (Matsumoto *et al.* 1982) create hot zones in the CVD gas phase either by a hot graphite disk (Spitsyn *et al.* 1981; Spitsyn 1991) or by means of a hot tungsten, molybdenum or tantalum wire (Matsumoto *et al.* 1982; Sawabe & Inuzuka 1985; Hirose & Terasaki 1986; Sommer & Smith 1991). For CTR, this 'hot zone' is at approximately 2000 °C (Spitsyn 1991). The filament temperature in a 'hot filament' diamond CVD reactor is usually between 2000 °C and 2400 °C (Sommer & Smith 1991). The growth rates for both methods are in the order of 1 μm h^{-1}, i.e. one to two orders of magnitude higher than in earlier thermal CVD attempts to grow diamond. While for hot filament CVD the carbonaceous gas phase is provided by feeding a mixture of usually less than 2% methane in hydrogen into a reactor, CTR is less specific. It relies on the etching of the hot, solid carbon source in hydrogen, the gasification of carbon-containing material and the subsequent redeposition of carbon in the desired phase on to substrates that are kept below 1000 °C. The etching process depends, of course, on the temperature of the solid carbon source, but also on its porosity, crystallinity,

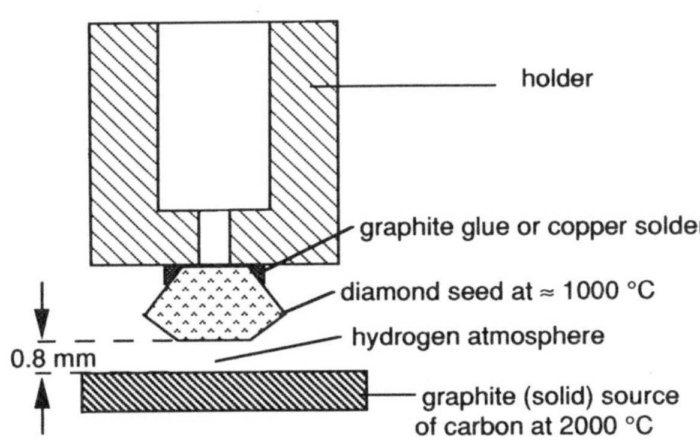

Figure 1. Homoepitaxial growth of diamond by means of 'chemical transport reaction' of carbon etched by hydrogen from a hot graphite disk (Spitsyn 1991).

size and shape. Figure 1 sketches the set-up of a CTR experiment to grow a homoepitaxial CVD diamond layer on a single crystal diamond substrate.

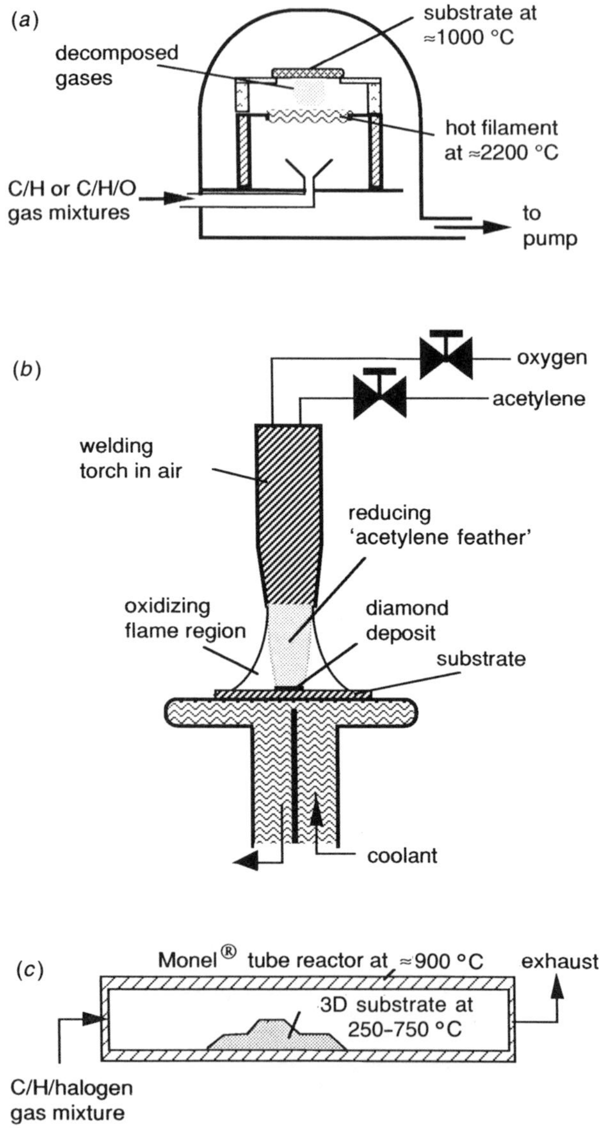

Figure 2. Operating principles of thermal diamond CVD methods: (*a*) hot filament reactor; (*b*) oxy-acetylene torch set-up; (*c*) Monel tube reactor for halogen-assisted CVD.

Another big step forward in terms of increasing the linear growth rates of diamond was the introduction of the oxy-acetylene torch synthesis, also a thermal CVD technique. This approach was first communicated by Hirose and co-workers (Hirose & Kondo 1988; Hirose *et al.* 1989; Matsui *et al.* 1991). They used a mixture of roughly 50% acetylene and 50% oxygen to grow diamond from an ordinary welding torch. They demonstrated that high linear diamond growth rates of 50–100 μm h^{-1} are achievable for hot flames that still show a reducing 'acetylene feather'. Gas phase temperatures of 3100 °C to 3300 °C are reported (Hirose & Kondo 1988). Meanwhile, Snail *et al.* (1991) and Janssen *et al.* (1990) were able to advance this method further and to demonstrate that high quality homoepitaxy at high rates is possible from oxy-acetylene flames.

Diamond single crystals at temperatures of up to 1300 °C were used as substrate material in such experiments. Recently, Doverspike and Butler (1993) were able to demonstrate the feasibility of a simple aerosol doping technique to incorporate boron into diamond films grown from such flames.

From an industrial point of view the flame synthesis of diamond seems less promising (with, maybe, the exception of increasing the weight of diamond seeds homoepitaxially). Temperature fluctuations that are difficult to control, small deposition areas and the enormous quantities of waste gases that, unlike with other methods, cannot be recycled into the reaction zone, limit the large scale application of combustion flame diamond deposition. The hot filament technique, on the other hand, has already found its way into the production facilities of several enterprises. The experimental set-up is simple and can be scaled to larger deposition areas by using a wire mesh rather than a single wire, and even three-dimensional substrates can be coated if the shape of the filament(s) mirrors the contours of the object to be coated. The major disadvantages of this method are the limited geometrical stability of the filament and its lifetime, contamination of the growing diamond by vaporized filament material and limitations to the gas composition (use of halogens and oxygen) and to the filament temperature due to filament corrosion. Figure 2 provides sketches of the hot filament, combustion flame and halogen-assisted thermal diamond CVD reactors. In tables 2 and 3, typical deposition parameters for hot filament CVD and combustion flame synthesis of diamond are presented.

Table 2. *Data for hot filament CVD of diamond*

substrate temperature	300–1000 °C
typical gas mixture	1% CH_4 in H_2; *small* quantities of oxygen tolerable
typical total gas flow	500–1000 sccm
temperature of the hot zone	2000–2400 °C (filament temperature)
deposition area	900 cm^2 demonstrated only limited by filament size and shape
linear growth rates	0.3–8 μm h^{-1}
advantages	simple set-up large deposition area demonstrated coating of 3D objects demonstrated
drawbacks	contamination of deposit by filament material bad filament stability and corrosion inhomogeneous deposition (moving substrates required for large areas) typical rates still fairly low only limited oxygen or halogen partial pressures tolerated

Table 3. *Data for combustion flame synthesis of diamond*

substrate temperature	600–1400 °C
typical gas mixture	51% acetylene and 49% O_2; hydrogen addition, ethylene or ethane feasible (lower rates)
typical total gas flow	1000–5000 sccm
temperature of the hot zone	3000–3200 °C (flame temperature)
deposition area	<1 cm^2
linear growth rates	30–200 μm h^{-1}
advantages	simple set-up high linear growth rates high quality homoepitaxy feasible
drawbacks	small deposition area inhomogeneous deposition process control difficult (T fluctuations, conc.) large quantities of waste gas contamination of deposit by torch material
additional remarks	flat flame burners allow for slightly larger deposition areas (2–3 cm^2) moving substrates are also feasible

(b) Plasma CVD *of diamond*

*(i) Low and medium pressure direct current (*DC*) plasma* CVD

Spitsyn *et al.* (1981) suggested the use of electrical discharges as a means to radicalize hydrogen and to decompose the carbon carrier gas in diamond CVD. A DC plasma discharge is probably the simplest way of forming an electrical discharge at low pressures. A number of papers covering DC plasma CVD of diamond have been published (Pinneo 1987; Ravi & Landstrass 1989; Thorsheim *et al.* 1991; Suzuki *et al.* 1987, 1990; Singh *et al.* 1988; Ohtake *et al.* 1989; Matsumoto *et al.* 1987; Kurihara *et al.* 1988; Matsumoto 1988; Bachmann *et al.* 1990; Lu & Bigelow 1992) and figure 3 depicts, as an example, the deposition system used by Suzuki *et al.* (1987, 1990). It consists of an anode, a cathode and the substrate mounted on to the anode. At low power levels and pressures, i.e. at low power densities, the gas temperature in such a system remains fairly low. Heat transfer from the discharge to the substrate is not sufficient to reach the desired substrate temperatures of 600–1000 °C. In this case, either the whole set-up is mounted inside a furnace to provide additional heating or the substrate is additionally heated.

At higher plasma power levels, heat transferred from the discharge is sufficient to heat the substrate and a further increase of the plasma power density requires cooling of the substrate. The substrate needs to be placed on to the anode of the reactor. If mounted on to the cathode, amorphous carbon rather than diamond is preferentially formed. Continuous, uniform coatings on wafers of more than 4 inches (10 cm) in diameter have been reported (Pinneo 1987; Ravi & Landstrass 1989) for low pressure/low power DC plasma CVD of diamond. The quality of the 'as-deposited' diamond is, however, only mediocre. It can take up substantial amounts of hydrogen that deteriorate the electrical properties of the material (Ravi & Landstrass 1989). Under such conditions, the deposition rate of less than 0.1 μm h^{-1} limit the industrial use of this technique. Both the quality of the deposit as well as the linear deposition rates are substantially improved if the reactor pressure is increased to approximately 200 mbar (Suzuki *et al.* 1987, 1990). In order to maintain a plasma at such pressures, it is necessary to increase the power fed into the system by increasing both the current density and the discharge voltage. These higher power densities result in much

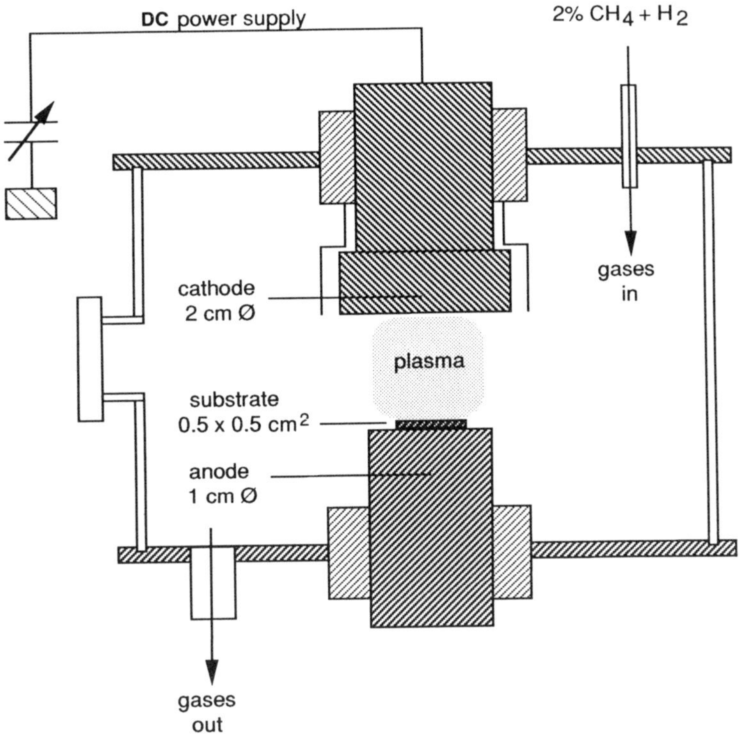

Figure 3. DC plasma CVD reactor used by Suzuki *et al.* (1987, 1990) for diamond deposition at different plasma power levels.

higher gas temperatures with the plasma being closer to local thermal equilibrium (LTE-plasma, hot plasma). The reactor set-up (figure 3) and the results published by Suzuki *et al.* (1987, 1990) are of specific interest, because this group used the same deposition set-up at different plasma power levels. More specifically, they increased the DC discharge current density from less than 1 A cm^{-2} to more than 10 A cm^{-2} and increased the linear growth rates from less than 20 μm h^{-1} to more than 250 μm h^{-1}. Moreover, they achieved these rates with a surprisingly low total gas throughput of only 20–100 sccm (std cm^3 min^{-1}) and without particularly directing any of the reactants towards the substrate surface or creating a turbulent gas phase (see figure 3). They measured the gas temperature spectroscopically and found it to be close to 4000 °C for a current density of 0.5 A cm^{-2}, 5300 °C for 4 A cm^{-2} and 6000 °C for 10 A cm^{-2}. Obviously, there is a trend that higher gas temperatures allow for higher growth rates and more effective conversion of the supplied carbon into diamond. At high power levels a quite remarkable amount of more than 40% of the carbon fed into the system is deposited as diamond (Suzuki *et al.* 1990). For methods with lower gas temperatures, e.g. hot filament CVD (2400 °C) or flame synthesis (3200 °C), less than 10% of carbon conversion and much lower linear growth rates are common. A major disadvantage

Table 4. *Data for low and medium pressure DC plasma CVD of diamond*

substrate temperature	600–1100 °C (cooling required for high pressure/power levels)
typical gas mixture	<2% CH_4 in H_2; oxygen no problem
typical total gas flow	100–500 sccm at low pressures 20–100 sccm at pressures of 100–200 mbar
temperature of the hot zone	<1400 °C at low pressures/power >6000 °C at 200 mbar and 10 A cm^{-2} (measured!)
deposition area	70 cm^2 at low pressure/power 1–2 cm^2 at 200 mbar and high plasma power
linear growth rates	<<0.1 μm h^{-1} (low pressure/power) 20–250 μm h^{-1} at 200 bar and 0.5–10 A cm^{-2}
advantages	large area at low pressure/power high linear growth rates at high pressure/power set-up fairly simple high carbon conversion rate (>40%) demonstrated at high power/pressure
drawbacks	bad quality at low pressure/power (H take-up) low rates at low pressure/power small deposition area at high pressure/power
additional remarks	annealing helps to remove hydrogen from deposits grown at low pressure/power substrates need to be mounted on to the system anode to avoid deposition of graphitic material

of the DC discharge plasma CVD of diamond at higher pressures over its low pressure variant is its severely reduced deposition area. Typical conditions and the results for DC plasma diamond deposition are summarized in table 4.

(ii) Diamonds deposited from DC arcs and DC jets

A further increase of the pressure in a DC plasma system leads to a deposition approach first introduced by Matsumoto and researchers at Fujitsu Laboratories (Matsumoto *et al.* 1987; Kurihara *et al.* 1988; Matsumoto 1988). They fed methane, hydrogen and argon into a conventional plasma torch commonly used for plasma spraying. Bachmann *et al.* (1990) used a cascaded arc plasma generator as a plasma source. In all cases, the gas phase inside the plasma generator is very hot. Temperatures of more than 5000 °C are easily obtained. The plasma is blown out of the generator into a reactor vessel thus forming a hot lance that creates a deposition zone of 1–2 cm^2. The basic principles of this approach are illustrated in figure 4.

The flame-like plasma expands from the plasma generator nozzle and, depending on the distance, may heat the deposition zone intensively. Therefore, the substrate usually needs to be cooled. The deposition results depend on a number of parameters, including the distance between the nozzle and the substrate surface, the gas velocity or the pressure difference between the generator and the reactor chamber. In those cases where the plasma expands as a supersonic plasma jet, the initially very hot gases inside the generator may cool down substantially, thus influencing the result of the deposition process. Bachmann *et al.* (1990) have shown that by simply varying the reactor pressure from 1 to 40 mbar, with the pressure inside the generator, the plasma power, the gas composition and the geometry of the set-up kept constant, the structure of the deposit changes from glassy carbon to polycrystalline diamond. Nevertheless, with DC plasma arc discharges at pressures between 0.2 and 1 bar, the highest diamond

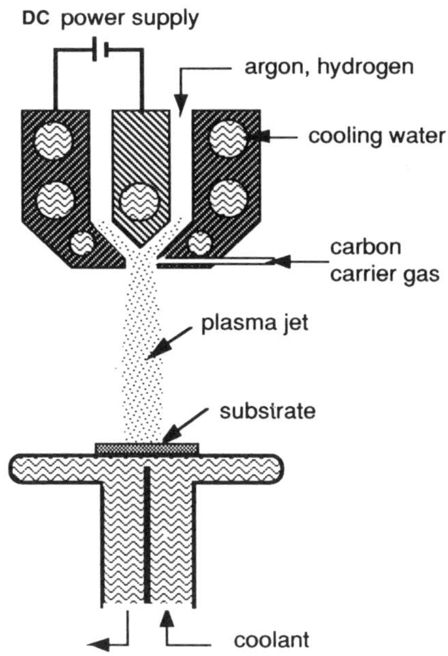

Figure 4. Utilization of a DC plasma jet for diamond deposition.

deposition rates for any method used to date were achieved. The top value of 930 μm h^{-1} has been reported by Ohtake *et al.* (1989). Interestingly, this research group did not even directly activate the carbon carrier gas. Rather than injecting methane into the plasma generator, they only passed hydrogen and argon through the generator. The hot plasma is blown out into the reactor vessel and methane is supplied elsewhere into the reactor vessel. The hot hydrogen/argon plasma alone provides sufficient energy to achieve high growth rates. The quality of plasma-jet deposited diamond can be extremely good, at least in the central portion of the deposition area. After polishing the rough surface of thick deposits, water-clear optically transparent diamond films are obtained. The major disadvantages of this approach are the relatively small deposition area, the non-uniformity in terms of the thickness as well as amorphous carbon contaminants at the periphery of the diamond deposit. In addition, the experimental set-up can be quite sophisticated and expensive and the stabilization of the plasma generation and substrate temperature needs considerable effort. Researchers at Norton Company, Northboro, Massachusetts (Lu & Bigelow 1992), recently at least partly overcame the restrictions of the deposition area. They used a magnetically stirred arc plasma generator of several hundred kilowatts in power. This unit allows the deposition of 250–300 μm thick films at a rate of approximately 15–25 μm h^{-1} over molybdenum substrates of 10 cm diameter. After deposition the diamond pops off the substrate owing to thermal expansion mismatch. The diamond disk is subsequently polished, laser-cut to size, metallized and is commercially available as thermal management substrates for heat dissipation in high power electronic circuitry. Typical data for the DC plasma jet deposition of diamond are listed in table 5.

Table 5. *Data for DC plasma jet CVD of diamond*

substrate temperature	800–1100 °C (cooling required)
typical gas mixture	<2% CH_4 in H_2; oxygen used to date
typical total gas flow	3000–70 000(!) sccm
temperature of the hot zone	>5000 °C
deposition area	2–3 cm^2
	70 cm^2 for high power magnetically stirred arc (Norton Co.)
linear growth rates	20–1000 μm h^{-1}, depending on set-up
	good diamond quality only at P >40 mbar
advantages	high linear growth rates at high power levels
	excellent quality at high power levels
	highest rates achieved to date
	high rate/large area combination demonstrated
drawbacks	small deposition areas for simple torches
	equipment can be sophisticated and expensive
	process control difficult
	bad film adhesion at high deposition rates
	high power consumption
	contamination of the deposit by electrode materials
	variable thickness and quality across the deposit
additional remarks	method is most suitable for fabrication of bulk diamond.
	well-adhering coatings on a variety of substrates is more difficult

(iii) Radiofrequency plasma CVD of diamond

The major driving force for using radiofrequency (RF) plasmas at low pressures for the preparation of diamond coatings is the availability of fully developed equipment capable of coating large areas of a variety of substrates with a variety of coating materials. From the equipment point of view, low pressure RF plasma CVD would have been the most likely candidate for the scale-up and industrialization of diamond deposition. Both the commonly used parallel plate reactor and inductively coupled low pressure RF plasmas have been tested (Matsumoto 1985; Meyer *et al.* 1989; Wood *et al.* 1990; Rudder *et al.* 1991). Undoubtedly, diamond can be deposited from such plasmas, but the synthesis of high quality coatings at low pressure is difficult. At the low pressures needed to take advantage of this type of plasma generator, amorphous, so-called diamond-like carbon (DLC) rather than diamond is formed. The properties of DLC coatings are entirely different from those of genuine diamond. Depending on its preparation parameters, DLC is optically transparent and soft or quite hard and brownish black. Its hardness, the only reason for its confusing name ('diamond-like') is, despite this name, always lower than that of diamond. DLC is a separate class of materials. Its ratio of sp^2 to sp^3 hybridized carbon varies with a number of deposition parameters. It may be useful for some low wear/low friction applications but it should not be confused with diamond. If diamond forms at all from low pressure RF plasmas, the material is usually embedded in an amorphous matrix and the composite is of mediocre quality. Frequently, diamond particles rather than dense films form. In parallel plate reactors, the electrodes need to be made from graphite in order to avoid contamination by sputtered electrode material. Similar to low pressure DC plasmas, the substrate, has to be additionally heated in order to perform experiments at substrate temperatures of ≈ 900 °C. The low pressure plasma alone is not hot enough to heat the substrate.

More promising for diamond deposition, but, to date, not fully exploited to its potential, is the use of atmospheric pressure inductively coupled RF thermal plasma torches (Matsumoto *et al.* 1987; Owano *et al.* 1991; Kohzaki *et al.* 1993). Similar to DC plasmas, the higher reactor pressure requires higher RF power levels to sustain a plasma. Such plasmas are typically created by inductively coupling 3–30 MHz RF power of 40–80 kW into a water-cooled silica plasma generator tube. A sketch of the experimental set-up is depicted in figure 5. Similar gas mixtures as for the other diamond preparation techniques have been used. However, in such plasmas

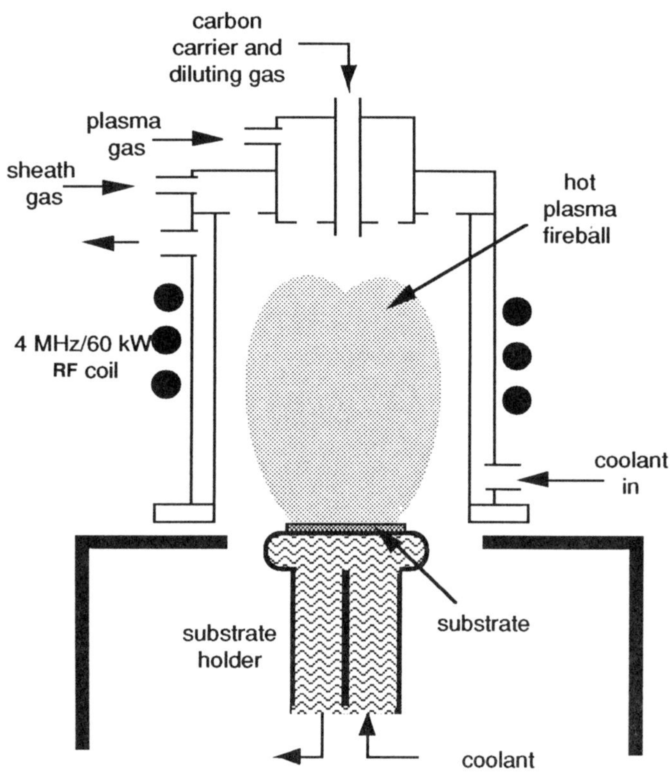

Figure 5. Inductively coupled, atmospheric pressure RF thermal plasma generator used by Matsumoto *et al.* (1987) for diamond deposition.

the gases are very hot and total gas flows as high as 80 000 sccm and special torch designs are common, in order to keep the plasma fireball away from the reactor walls.

At growth rates of up to 180 μm h^{-1}, high quality diamond films can be obtained. The deposition area is usually small, but comparable with those of medium pressure DC discharges or DC plasma jets. However, uniform diamond deposits of 10 cm diameter prepared by RF thermal plasma deposition of diamond coatings for tool applications were recently reported by Kohzaki *et al.* (1993). By inductively coupling the energy into the system, contamination of the deposit by plasma generator material seems to be avoidable. However, the handling of such plasmas is not easy. They are very hot and key problems with this technology are to avoid contact of the plasma fireball with the silica generator walls, position it correctly with respect to the substrate and stabilize the substrate temperature below 1100 °C. Other problems seem to be the adhesion of the film and the choice of substrates applicable in such a harsh environment. Uniformity of the deposit is usually a problem too, but the work of Kohzaki *et al.* indicates that this can be solved at least for deposits of 10 cm diameter. Data on RF thermal plasma diamond CVD are still limited and additional work is required to exploit fully the potential of this method. However, recent diamond-related conferences have shown considerable progress in this direction. The deposition conditions, results, advantages and drawbacks of low and atmospheric pressure RF plasma diamond CVD are outlined in table 6.

Table 6. *Data for* RF *plasma diamond deposition*

substrate temperature	700–1200 °C (cooling required for high pressure/power)
typical gas mixture	<2% CH_4 in H_2; oxygen not used to date
typical total gas flow	100–200 sccm for low pressure discharges 80 000–100 000 (!) sccm for atmospheric pressure RF torches
temperature of the hot zone	<1500 °C for typical low pressure discharges >5000 °C for high pressure torches
deposition area	large areas for low pressures (but bad quality) 2–3 cm^2 for torches
linear growth rates	<0.1 μm h^{-1} (low pressure) 30–180 μm h^{-1} (RF torch)
advantages	high linear growth rates at high power levels excellent quality at high power levels homogeneity over 10 cm diameter demonstrated
drawbacks	small deposition areas for simple torches equipment can be sophisticated and expensive process control difficult bad film adhesion at high deposition rates high power consumption bad quality (mixed phase material) at low pressures/power large amounts of sheath and diluting gas required
additional remarks	method is not yet fully exploited to its potential only inductively coupled RF plasma generators are suitable for diamond CVD

(iv) Microwave plasma CVD *of diamond*

The diamond research group at the Japanese National Institute for Research in Inorganic Materials (NIRIM) were the first to report the utilization of a 2.45 GHz microwave plasma for diamond deposition (Kamo *et al.* 1983). Without doubt, it is this and the 'hot filament' technique (invented by the same group) that moved diamond thin films much closer to an industrially applicable technology. Many of the laboratory set-ups (Badzian *et al.* 1986; Kobashi *et al.* 1988) and even a few production units in use (e.g. at Idemitsu Petrochemical; see the photograph in the review article by Bachmann & Messier 1989) are similar to the apparatus originally described by Kamo *et al.* (1983). Its cross-section is shown in figure 6, along with a sketch of a different design that was jointly developed by Bachmann and ASTeX, Applied Science and Technology, Woburn, Massachusetts (Bachmann *et al.* 1988a; and the review articles by Bachmann & Messier 1989; Bachmann *et al.* 1988; Bachmann 1991).

In the original NIRIM set-up, the plasma forms inside a silica tube. The substrate size is limited by the tube diameter and plasma etching of the reactor wall can cause severe contamination of the growing film. The bell jar system (Bachmann *et al.* 1988*a*) allows for up to 10 cm (4 inch) diameter substrates and greatly reduces contamination problems by utilizing a ball-shaped plasma that only touches the substrate surface. The substrates can be separately heated or cooled, depending on the microwave plasma power supplied. Jiang & Klages (1993) recently equipped such a unit with the capability of applying a bias voltage between substrate and plasma. This group showed, successfully, that the abrasive polishing procedure commonly used to enhance the nucleation density of diamond (Bachmann *et al.* 1988*a*) can be replaced by bias-enhanced predeposition from a carbon-enriched plasma. They also were able to show that such a pretreatment creates nucleation sites that are oriented with respect to the silicon substrate lattices and that diamond can grow heteroepitaxially from these nucleation sites. A number of similar reactor designs were subsequently tested by Kamo *et al.* (1990), Ishibori & Ohira (1990) and J. Asmussen of Wavemat Inc, Michigan. These set-ups include the use of several microwave sources fed into a reactor chamber (Kamo *et al.* 1990), the use of a microwave horn rather than a microwave antenna (Ishibori & Ohira 1990) or the utilization of a smaller silica cup to contain the plasma. The approaches described in Kamo *et al.* (1990) suffer from plasma instabilities, and that of Wavemat Inc. produces a plasma that again etches silica reactor walls. Therefore, despite all other attempts, varieties of the design described in Bachmann *et al.* (1988*a*) are probably the most common in today's microwave plasma deposition units. Nevertheless, this design has its limitations, too. One is the limited substrate area of 20–80 cm^2 (depending on reactor design), another is the uniformity of the coating. Utilization of 915 MHz microwaves rather than 2.45 GHz would allow for larger reactors and larger plasmas and, therefore, may help by-pass some of these limitations.

Microwaves of 2.45 GHz were also utilized by Mitsuda *et al.* (1989) in a system that operates at atmospheric pressure rather than at the 20–200 mbar operating regime of the other microwave plasma units. At higher pressures, as mentioned earlier, the plasma is closer to thermal equilibrium and allows for higher gas temperatures. Similar to the trend already observed when comparing low and high pressure DC or RF plasmas, this results in a substantial increase of the diamond growth rate. While at power levels of up to 1.5 kW linear growth rates of 0.5–3 $\mu m\ h^{-1}$ are common for both the tubular and the bell jar reactor, rates of up to 30 $\mu m\ h^{-1}$ were achieved in Mitsuda's microwave torch approach. A cross-section of this apparatus is also included

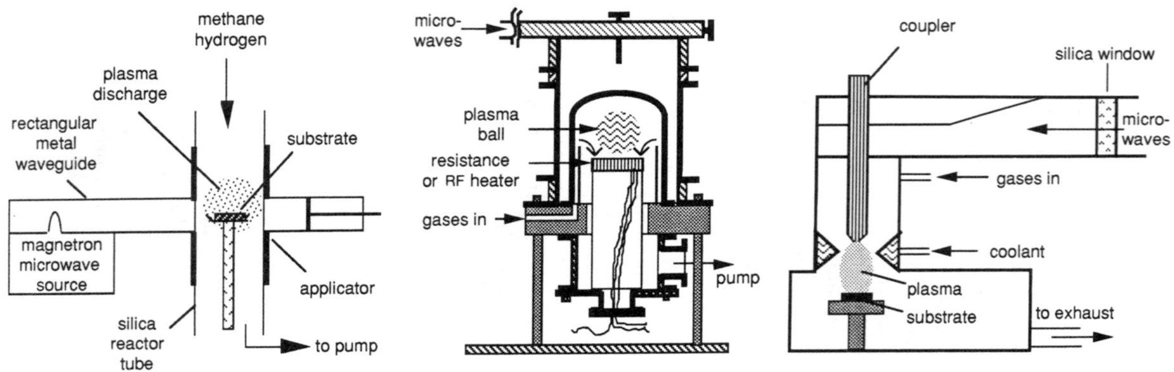

Figure 6. Microwave plasma CVD of diamond. Cross-sections of the tubular reactor developed by Kamo *et al.* (1983) (left), the bell jar reactor by Bachmann jointly with ASTeX (Bachmann *et al.* 1988), and of an atmospheric pressure microwave plasma torch set designed by Mitsuda *et al.* (1989) (right) are depicted.

Table 7. *Data for microwave plasma* CVD *of diamond*

substrate temperature	300–1200 °C (? plasma emissions a problem) (cooling required for high pressure/power)
typical gas mixture	all C/H/O mixtures according to figure 7 feasible C/H mixtures still dominate the literature
typical total gas flow	<1 sccm (nearly closed system) for solid C source hydrogen plasma reactor (CTR) 100–1000 sccm for standard C/H mixtures 5–50 sccm for C/H/O mixtures 50 000 (!) sccm for atmospheric pressure torch
temperature of the hot zone	≈ 2500 °C at 50 mbar/1.5 kW higher at higher pressure/power levels
deposition area	70–80 cm^2 at low pressures/powers 20 cm^2 for 100 mbar/3–4 kW; <5 cm^2 for torch
linear growth rates	0.5–15 $\mu m\ h^{-1}$, depending on pressure/power
advantages	reasonable linear growth rates and areas excellent quality of the material good control of the deposition parameters stable system operation adhering films on a wide variety of substrates C/H, C/H/O, C/H/halogen mixtures feasible
drawbacks	small deposition areas for torch areas and rates need improvement 3D substrates difficult
additional remarks	915 MHz and 8 GHz were also used magnetized (ECR) plasmas have no advantage magnetrons of more that 6 kW are difficult to buy rates and quality at low substrate temperatures are insufficient for practical use

in figure 5. In table 7, data are given for both the common low pressure 2.45 GHz microwave plasma CVD technique and the 2.45 GHz plasma torch experiments, indicating the advantage of the higher rates and the drawback of the smaller area for the latter set-up. By supplying more than 3 kW of microwave power to a variant of the bell jar reactor Bachmann *et al.* (1993) recently partly overcame these limitations and grew diamond homogeneously on to 16 cm^2 substrates at rates of more than 12 $\mu m\ h^{-1}$.

Magnetized and electron cyclotron resonance (ECR) 2.45 GHz microwave plasmas were also used for diamond deposition experiments (Kawarada *et al.* 1987; Suzuki 1989; Yuasa *et al.* 1991). At pressures of more than 10 mbar, the electron collision rate in the plasma is far too high for full ECR action. The plasma is hardly affected by the presence of a magnetic field and the deposition results are no different from those from non-magnetized microwave plasma CVD. At pressures of less than 0.1 mbar, where ECR starts to be significant, both the deposition rate and the crystal size are markedly reduced, compared with the conventional microwave plasma CVD of diamond, and the corresponding Raman data indicate substantial amounts of amorphous material. Although the electron temperature in such plasmas is still high, the gas temperature drops considerably under these low pressure conditions. The fact that only mixed phase material is deposited from low pressure ECR plasmas again underlines the dominant role of high gas (heavy particle) temperatures for diamond formation.

Microwave frequencies other than the common 2.45 GHz were also exploited, namely 0.915 GHz (Miyake *et al.* 1988) and X-band frequencies around 8 GHz (Aklufi

& Brock 1989). The X-band experiments were performed at rather low power levels of only 300 W and at pressures of approximately 0.1 mbar and only amorphous DLC films formed. The 915 MHz experiments by Miyake *et al.* (1988) were carried out at both 25 mbar and atmospheric pressure using the same experimental set-up, but with 1 kW and 7 kW input power, respectively. At 25 mbar, rates of approximately 1 $\mu m\ h^{-1}$, i.e. similar to the 2.45 GHz experiments, were achieved. At 1 bar, polycrystalline diamond grew at a rate of up to 50 $\mu m\ h^{-1}$. This correlates well with the results obtained at 2.45 GHz and illustrates that the excitation frequency of the plasma is not important for the deposition process. A pressure increase accompanied by a higher power input, i.e. higher power densities and, consequently, higher gas temperatures, again result in higher deposition rates.

*(c) Non-*CVD *methods of forming diamond at low pressure*

In addition to the various CVD routes to grow diamond at low pressures, a number of non-CVD approaches, i.e. approaches where chemical reactions are not involved in the deposition process, exist. This implies that the starting material is solid carbon and that, other than in CTR deposition, gasification, formation of a carbon-containing carrier gas and its subsequent decomposition are not involved. To date, all approaches to growing diamond using solid carbon as a source material either are modifications of CTR, i.e. the generation of hydrocarbons is an important intermediate step (Roy 1993; Bachmann 1993), or do not result in diamond formation. Physical vapour deposition (PVD) such as the sputtering of graphite targets or the laser ablation of graphite, as well as carbon ion beam techniques, have not been capable of depositing diamond. Sometimes such deposits are, for whatever reason, termed 'amorphous diamond', which is a clear contradiction in terms. This material may be interesting and useful for certain applications. The same holds for 'diamond-like' carbon (DLC). However, owing to a lack of long range periodicity, it is definitely *not* diamond: it has neither the crystal structure of diamond nor its band gap nor its thermal conductivity nor any other characteristic diamond properties.

Prins & Gaigher (1991) recently showed that the implantation of carbon into non-carbide-forming materials such as copper and the subsequent treatment of the implanted surface by an energetic beam, e.g. by a high power laser or an ion beam, can result in the formation of crystalline diamond particles. However, not all attempts to repeat these experiments were successful (Lau *et al.* 1992).

Another non-CVD diamond formation method was first communicated by Fedoseev *et al.* (1983) and later confirmed by Aslam *et al.* (1989). Small soot particles are rapidly heated in a laser beam and, by creating a high pressure level inside the little carbon spheres, part of the material is (high pressure) converted into diamond.

Scientifically the non-CVD methods of forming diamond are very interesting; however, they are far from any practical application and, hence, are not discussed in any further detail.

(d) The choice of diamond CVD *method*

To conclude the comparison of methods, the present status of diamond CVD is summarized in table 8. It illustrates that a variety of different methods are capable of depositing high quality coatings on to various non-diamond substrates. Some of these methods, like microwave plasma CVD (used for production at, for example, De Beers, U.K., Idemitsu Petrochemical, Japan), hot filament CVD (used for production at, for example, Diamonex Inc., U.S.A.) or DC jet CVD (used for production at Norton

Table 8. *Present status of low pressure diamond formation methods* (1 atm ≈ 10^5 Pa.)

method	rate/(μm h^{-1})	area/ (cm^2)	quality	advantages	drawbacks
combustion flame	30–200	0.5–3	+++ (homoepi.)	simple	area, stability
hot filament	0.3–8	5–900	+++	simple large area	contaminations filament stability
DC discharge (low pressure)	<0.1	70	−/+ (annealing)	simple large area	quality, rate
DC discharge (medium pressure)	20–250	<2	+++	high rate excellent quality	area
DC plasma jet	930 (25)	<2 (100)	+++	highest rate excellent quality	contaminations, stability, equipment costs
RF (low pressure)	<0.1	100	−	scale-up	quality, rate (contamination)
RF (thermal, 1 atm)	180 (30)	2–3 (70)	+++	rate quality	area, stability, equipment costs
microwave (0.9–2.45 GHz)	1 (low pressure) 30 (high pressure)	80 5	+++	quality stability	rate, area
microwave (ECR 2.45 GHz)	0.1	100	−/+	area	quality, rate (contamination)

Company, U.S.A.), are already well advanced; however, all methods could definitely profit from further scale-up. Which method to choose depends primarily on the specific application intended for the diamond film. If contamination-free, uniform coatings on medium size substrates are required at fairly low equipment costs, microwave plasma CVD is probably a good choice. Such units also have the advantage of running quite stably over extended periods of time and the deposition process can be automated. For bulk material, DC arc discharges or DC jets seem to be most promising. In those cases where contamination of the film by the filament material is not a drawback and the coating of large areas or of three-dimensional substrates is required, the hot filament method certainly has its advantages. The state of the art of all other methods is presently far from the level needed in an industrial environment.

3. Correlations and general aspects

From table 8, the wide variety of available diamond formation methods is quite obvious. These methods operate under very different conditions, start from a confusing variety of precursor materials, and are applied to a wide range of substrate materials. The achieved rates, quality and phase purity of the deposit vary considerably. Recently, Bachmann *et al.* (1991, 1993) made an effort to develop a common scheme to explain at least some of these differences. Their data analysis covers 30 years of CVD diamond research and more than 80 different experiments. Part of the results of this analysis is shown in figure 7. In a C/H/O gas phase compositional diagram, the successful diamond deposition experiments (indicated by a diamond symbol) were found to be located in a concentration field in the centre of the ternary diagram, in a ‘diamond

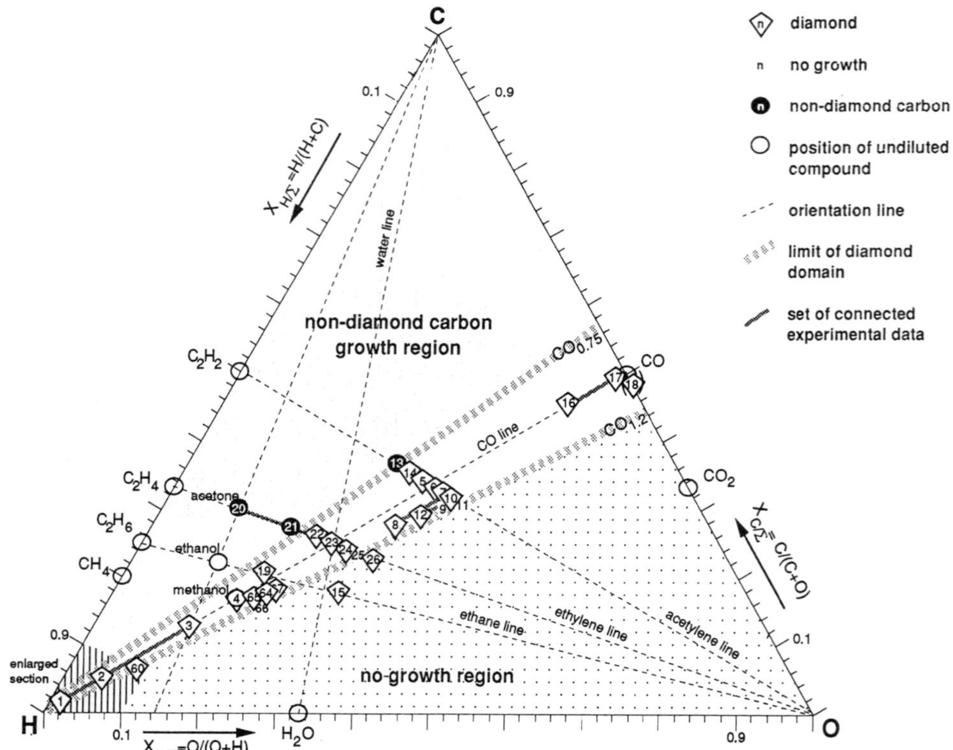

Figure 7. The C/H/O diagram of diamond CVD (Bachmann *et al.* 1991) reveals that diamond deposition is only feasible in a narrow diamond domain in the centre of this gas phase compositional diagram. This diagram includes datapoints from more than 30 sources, obtained in over 80 experiments and relating to thermal CVD, hot filament data, microwave plasma (MW) plasma CVD, DC jet data, RF torch CVD, ECR MW plasma CVD, DC glow discharge data, flames.

domain', positioned along the CO–H line of the diagram. On the oxygen-rich side, the diamond domain is limited by a region where no material at all is deposited, but on the carbon-rich side, deposition of non-diamond carbon dominates. Thus, as a general trend, the phase purity of the material increases from the carbon-rich to the oxygen-rich side of the diamond domain. For oxygen partial pressures that are too high, however, no solid carbon forms. This also implies that the deposition rates have to decrease when traversing the diamond domain from its carbon-rich to its oxygen-rich border line. This diagram has been confirmed with slight modifications (Bachmann *et al.* 1993) in many additional experiments and provides a helpful guideline as well as a strategy tool to design and optimize diamond CVD experiments. However, the presence of a diamond domain does not imply that under all circumstances diamond has to condense from a gas mixture with just the correct composition. Rather, it means that it is worth while to search for deposition conditions like temperature, plasma power, substrate pretreatments, etc., within the diamond domain as a limit for the gas composition. Halogen-containing gases are not included in this diagram, but C/H/F or C/H/Cl diagrams have already been suggested (Bachmann *et al.* 1991) for such mixtures. Ternary gas mixtures have certain advantages over the binary C/H mixtures (Bachmann *et al.* 1993), e.g. lower deposition temperatures. Whether or not quaternary gas mixtures have any advantage over ternary mixtures remains to be seen, but if so, a three-dimensional gas phase compositional diagram, e.g. a C/H/O/halogen tetrahedron, would be the appropriate way of representing them.

The rate variations across the diamond domain alone are not sufficient to explain the huge rate differences for the various deposition methods. Another explanation is required. In diamond CVD, the substrate temperature has always to remain below 1400 °C to prevent graphitization of the growing film, i.e. for high rate synthesis the gas phase is always rapidly quenched from very high to relatively moderate, similar substrate temperatures. Therefore, the substrate temperature is not the decisive parameter. Other deposition parameters such as a specific way of initiating the CVD processes or the excitation frequency to sustain a plasma are seemingly also not important. Rates of less than 0.1 and 1000 $\mu m\ h^{-1}$ were reported for DC plasmas and rates of less than 0.1 and 180 $\mu m\ h^{-1}$ were reported for RF discharges. High rates of 100 $\mu m\ h^{-1}$ were reported for combustion flames, i.e. not a plasma CVD method but a thermal CVD method. In several cases, high deposition rates seem to correlate with high total gas flows. However, Suzuki *et al.* (1987, 1990) have demonstrated that even at total gas flows of only 20–100 sccm growth rates of 20–250 $\mu m\ h^{-1}$ are feasible. In their experiments, variations of the plasma power and hence the temperature of the CVD gas phase were sufficient to change the rates considerably. From a general analysis, Bachmann *et al.* (1991) concluded (see also the review article by Bachmann & Lydtin 1990) that a hot spot in the CVD gas phase with gas temperatures of more than 4000 °C is desirable for high rate diamond synthesis. The generation of carbon vapour may be an important intermediate step to produce the molecular fragment efficiently, e.g. methyl or CH radicals that finally are incorporated into the diamond surface. Of course, conservation of these species on their way to the substrate is necessary and can be achieved by extending the plasma to the very surface of the substrate or by very fast transport of such species from the source to the surface. The correlation between the linear growth rates and the approximate temperatures of the CVD gas phases is depicted in figure 8.

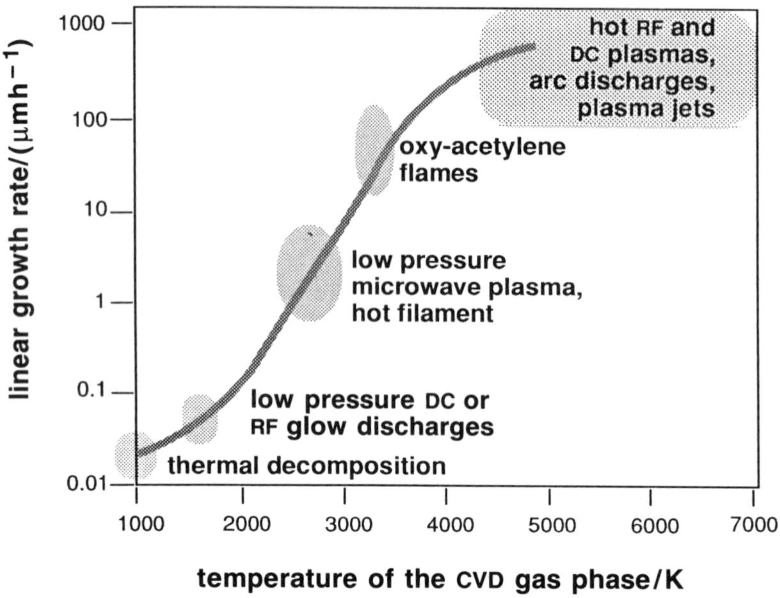

Figure 8 Correlation between the temperaure of the CVD gas phase and the reported linear growth rates of currently used diamond deposition techniques (Bachmann *et al.* 1991 and the review article by Bachmann & Lydtin 1990).

4. Summary and outlook

During the past decade, diamond thin film deposition technologies have come a long way. Deposition rates have increased from less than 0.1 $\mu m\ h^{-1}$ to almost 1000 $\mu m\ h^{-1}$. Deposition areas went from 1 cm^2 to more than 200 cm^2 and substrate temperatures dropped from 900 °C to less than 500 °C, or even lower. These results were mostly achieved in separate experiments. However, for many applications, diamond needs to be deposited: at high deposition rates; on large area substrates; with complex shapes; and at low substrate temperatures. In addition, the material has to stick to the substrate surface and quite often a certain surface morphology is also required. This means that the 'champion' data of one or another method now have to be integrated into a single experiment. Reproducibility and uniformity of the film properties, scale-up of the reactors, and doping of CVD diamond are other important deposition-technology-related issues that need further attention and work. Despite these problems and with the first CVD diamond products already on the market, as well as several products and devices at least demonstrated as prototypes, there is no doubt that the emerging technology of diamond thin films will play a major role in the field of advanced materials.

This work is partly being sponsored by the German Ministry of Research and Technology. The author is indebted to D. Leers, D.U. Wiechert, H. Lade, H. Lydtin and many other colleagues within the Philips Research Organization for their active support, valuable discussions and constant encouragement.

References

Aklufi, M. & Brock, D. 1989 *Proc. First Int. Symp. on Diamond and Diamond-Like Films*, vol. 89-12, p. 114. Pennington, New Jersey: The Electrochemical Society.

Angus, J. C., Will, H. A. & Stanko, W. S. 1968 *J. appl. Phys.* **39**, 2915.

Aslam, M., Debroy, T., Roy, R. & Breval, E. 1989 *Carbon* **27**, 289.

Bachmann, P. K. 1993 *Adv. Mater.* no. 2/93, 137.

Bachmann, P. K., Drawl, W., Knight, D., Weimer, R. & Messier, R. F. 1988a In *Diamond and diamond-like materials* (ed. A. Badzian, M. Geis & G. Johnson), extended abstracts, vol. EA-15, p. 99. Pittsburgh: Materials Research Society.

Bachmann, P. K., Weimer, R. & Messier, R. 1988b *Diamond Technology Initiative Symp. 12–14 July 1988, Crystal City, Arlington, Virginia*, technical digest, paper T2.

Bachmann, P.K., Lydtin, H., Wiechert, D. U., Beulens, J. J., Kroesen, G. & Schram, D. C. 1990 In *Surface modification technologies III* (ed. T.S. Sudarshan & D. G. Bhat), p. 69. Warrendale, Pennsylvania: The Minerals, Metals & Materials Society (TMS).

Bachmann, P. K., Leers, D. & Lydtin, H. 1991 *Diamond Related Mater.* **1**, 1.

Bachmann, P. K., Leers, D. & Wiechert, D. U. 1993 *DIAMOND 1992, 3rd Eur. Conf. on Diamond, Diamond-Like and Related Materials jointly with 3rd Int. Conf. on the New Diamond Science and Technology, 31 Aug.–4 Sept. 1992, Heidelberg, Germany, paper 4.2.* (Full paper in preparation, 1993.)

Badzian, A. R., Simonton, B., Badzian, T., Messier, R., Spear, K. E. & Roy, R. 1986 *Proc. SPIE* **683**, 127.

Chang, C. P., Flamm, D. L., Ibbotson, D. E. & Mucha, J. A. 1988 *J. appl. Phys.* **63**, 1744.

Doverspike, K. & Butler, J. E. 1993 In *DIAMOND 1992* (ed. P. K. Bachmann, A. T. Collins & M. Seal), p. 1078. Lausanne: Elsevier Sequoia.

Eversole, W. 1962 U.S. Patents 3,030,187 and 3,030,188 (filed in 1959).

Fedoseev, D. V., Bukhovets, V. L. & Varshavskaya, I.G. 1983 *Carbon* **21**, 237.

Hibshman, H. J. 1968 U.S. Patent 3,371,996 (filed in 1964).

Hirose, Y. & Kondo, N. 1988 *Extended Abstracts, 35th Spring Meeting, Jap. Appl. Phys. Soc. 29 March*, p. 434.

Hirose, Y. & Terasaki, Y. 1986 *Jap. J. appl. Phys.* **25**, L519.

Hirose, Y., Ananuma, S., Okada, N. & Komaki, I. 1989. *Proc. First Int. Symp. on Diamond and Diamond-Like Films*, vol. 89-12, p. 80. Pennington, New Jersey: The Electrochemical Society (1989).

Ishibori, K. & Ohira, Y. 1990 In *Science and technology of new diamond* (ed. S. Saito, O. Fukanaga & M. Yoshikawa), p. 167. Tokyo: KTK Science Publishers.

Janssen, G., van Enckefort, W. J. P., Schaminee, J. J. D., Vollenberg, W., Giling, L. J. & Seal, M. 1990 *J. Cryst. Growth* **104**, 752.

Jiang, X. & Klages, C. P. 1993 In *DIAMOND 1992* (ed. P. K. Bachmann, A. T. Collins & M. Seal), p. 1112. Lausanne: Elsevier Sequoia.

Kamo, M., Sato, Y., Matsumoto, S. & Setaka, N. 1983 *J. Cryst. Growth* **62**, 642.

Kamo, M., Takamura, F. & Sato, Y. 1990 In *Science and technology of new diamond* (ed. S. Saito, O. Fukanaga & M. Yoshikawa), p. 183. Tokyo: KTK Science Publishers.

Kawarada, H., Mar, K. S. & Hiraki, A. 1987 *Jap. J. appl. Phys.* **26**, 6, L1032.

Kobashi, K., Nishimura, K., Kawate, Y. & Horiuchi, T. 1988 *Phys. Rev.* B **38**, 4067.

Kohzaki, M., Higuchi, K., Noda, S. & Uchida, K. 1993 In *DIAMOND 1992* (ed. P. K. Bachmann, A. T. Collins & M. Seal), p. 612. Lausanne: Elsevier Sequoia.

Kurihara, K., Sasaki, K., Kawarada, M. & Koshino, N. 1988 *Appl. Phys. Lett.*, **52**, 6, 437.

Lau, W. M., Feng, X., Bello, I., Lee, S. T., Chen, S. & Braunstein, G. 1992 In *Diamond, diamond-like and related films 1991* (ed. P. K. Bachmann & A. Matthews). Lausanne: Elsevier Sequoia.

Liou, Y., Weimer, R., Knight, D. & Messier, R. 1990 *Appl. Phys. Lett.* **56**, 437.

Lu, G. & Bigelow, L.K. 1992 In *Diamond, diamond-like and related films 1991* (ed. P. K. Bachmann & A. Matthews), p. 1064. Lausanne: Elsevier Sequoia.

Matsui, Y., Yabe, H., Sugimoto, T. & Hirose, Y. 1991 In *Diamond and diamond-like carbon coatings* (ed. A. Matthews & P. K. Bachmann). Lausanne: Elsevier Sequoia.

Matsumoto, S. 1985 *J. Mater. Sci. Lett.* **4**, 600.

Matsumoto, S. 1988 In *Diamond and diamond-like materials synthesis* (ed. G.H. Johnson, A.R. Badzian

& M. W. Geis), *Mater. Res. Soc. Symp.* extended abstracts, EA-15, p. 119, Pittsburgh.
Matsumoto, S., Sato, Y., Kamo, M. & Setaka, N. 1982 *Jap. J. appl. Phys.* **21**, L183.
Matsumoto, S., Hino, M. & Kobayashi, T. 1987 *Appl. Phys. Lett.* **51**, 10, 737.
Matsumoto, S., Hino, M., Moriyoshi, Y., Nagashima, T. & Tsutsmui, M. 1988 U.S. Patent 4,767,608 (filed 19 Oct. 1987).
Meyer, D. E., Dillon, R. O. & Woolham, J. A. 1989 *J. Vac. Sci. Technol.* **7**, 3, 2325.
Mitsuda, Y., Yoshida, T. & Akashi, K. 1989 *Rev. Sci. Instrum.* **60**, 2, 249.
Miyake, S., Chen, W., Hoshino, A. & Arata, Y. 1988 *Trans. JWRI*, **17**, 2, 323.
Ohtake, N., Tokura, H., Kuriyama, Y., Mashimo, Y. & Yoshikawa, M. 1989 *Proc. First Int. Symp. on Diamond and Diamond-Like Films*, vol. 89-12, p. 93. Pennington, New Jersey: The Electrochemical Society.
Owano, T. G., Goodwin, D. G., Kruger, C. H. & Cappelli, M. A. 1991 In *New diamond science and technology* (ed. R. F. Messier, J. T. Glass, J. E. Butler & R. Roy), p. 497. Pittsburgh: Materials Research Society.
Patterson, D. E., Bai, B. J., Chu, C. J., Hauge, R. H. & Margrave, J. L. 1991 In *New diamond science and technology* (ed. R. F. Messier, J. T. Glass, J. E. Butler & R. Roy), p. 433. Pittsburgh: Materials Research Society.
Pinneo, J. M. 1987 *1st Diamond Technology Initiative Workshop, MIT Lincoln Laboratories, Boston, 2 Feb.*, paper 4.
Prins, J. & Gaigher, H. L. 1991 In *New diamond science and technology* (ed. R. F. Messier, J. T. Glass, J. E. Butler & R. Roy), p. 561. Pittsburgh: Materials Research Society.
Ravi, K. V. & Landstrass, M. I. 1989 *Proc. First Int. Symp. on Diamond and Diamond-Like Films*, vol. 89-12, p. 24. Pennington, New Jersey: The Electrochemical Society.
Roy, R., 1993 *Adv. Mater.* no. 2/93, 141.
Rudder, R. A., Posthill, J. B., Hudson, G. C., Malta, D., Thomas, R. E., Markunas, R. J., Humphrey, T. P. & Nemanich, R. J. 1991 In *New diamond science and technology* (ed. R. F. Messier, J. T. Glass, J. E. Butler & R. Roy), p. 425. Pittsburgh: Materials Research Society.
Sawabe, A. & Inuzuka, T. 1985 *Appl. Phys. Lett.* **46**, 2, 146.
Singh, B., Mesker, O. R., Levine, A. W. & Arie, Y. 1988 *Appl. Phys. Lett.* **52**, 1658.
Smith, D. K., Sevillano, E., Besen, M., Berkman, V. & Bourget, L. 1992 In *Diamond, diamond-like and related films 1991* (ed. P. K. Bachmann & A. Matthews). Lausanne: Elsevier Sequoia.
Snail, K., Oakes, D., Butler, J. & Hanssen, L. 1991 In *New diamond science and technology* (ed. R. F. Messier, J. T. Glass, J. E. Butler & R. Roy), p. 503. Pittsburgh: Materials Research Society.
Sommer, M. & Smith, F. W. 1991 In *New diamond science and technology* (ed. R. F.Messier, J. T. Glass, J. E. Butler & R. Roy), p. 443. Pittsburgh: Materials Research Society.
Spitsyn, B. V. & Derjaguin, B. V. 1956 Author's certificate (patent application), 10 July. U.S.S.R. Patent 339,134.
Spitsyn, B. V., Bouilov, L. L. & Derjaguin, B. V. 1981 *J. Cryst. Growth*, **52**, 219.
Spitsyn, B.V. 1991 In *Applications of diamond films and related materials* (ed. Y. Tzeng, M. Yoshikawa, M. Muranaka & A. Feldman), p. 475. Amsterdam: Elsevier.
Suzuki, J. 1989 *Jap. J. appl. Phys.* **28**, 2, L281.
Suzuki, K., Sawabe, A., Yasuda, H. & Inuzuka, T. 1987 *Appl. Phys. Lett.* **50**, 12, 728.
Suzuki, K., Sawabe, A., Yasuda, H. & Inuzuka, T. 1990 *Jap. J. appl. Phys.* **29**, 153.
Thorsheim, H. R., Celii, F. G., Butler, J. E., Plano, L. S. & Pinneo, J. M. 1991 In *New diamond science and technology* (ed. R. F. Messier, J. T. Glass, J. E. Butler & R. Roy), p. 207. Pittsburgh: Materials Research Society.
Wood, P., Wydeven, T. & Tsuji, O. 1990 In *Science and Technology of New Diamond* (ed. S. Saito, O. Fukanaga & M. Yoshikawa), p. 167. Tokyo: KTK Science Publishers.
Yuasa, M., Kawarada, H., Wei, J., Ma, J. S., Suzuki, J., Okada, S. & Hiraki, A. 1991. *Surf. Coating Technol.* **49**, 374.

Selected review articles

Angus, J. C. & Hayman, C. C. 1988 *Science, Wash.* **241**, 913.
Bachmann, P. K., 1991 *Ber. Bunsenges, Phys. Chem.* **95**, 1390.

Bachmann, P. K. & Lydtin, H. 1990 In *Characterization of plasma processes* (ed. G. Lukovsky, D.E. Ibbotson & D. W. Hess), vol. 165, p. 181. Materials Research Society: Pittsburgh.
Bachmann, P. K. & Messier, R. F. 1989 *Chem. Eng. News* **67**, 20, May 15, 24.
Bachmann, P. K. & van Enckefort, W. J. P. 1992 *Diamond Related Mater.* **1**, 1024.
Bachmann, P. K., Gärtner, G. & Lydtin, H. 1988 *MRS Bull.* **12**, 52.
Badzian, A. R. & DeVries, R. C. 1988 *Mater. Res. Bull.* **23**, 385.
DeVries, R. C. 1987 *A. Rev. Mater. Sci.*, **17**, 161.
Lux, B. & Haubner, R. 1989 *Proc. 12th Int. Plansee-Seminar*, p. 615.
Spear, K. 1989 *J. Am. Ceram. Soc.* **72**, 171.
Spitzyn, B. V. 1990 *J. Cryst. Growth* **99**, 1162.
van Enckevort, W. J. P. 1990 *J. Hard Mater.* **1**, (4).

4

Local epitaxial growth of diamond on nickel from the vapour phase

BY Y. SATO, H. FUJITA, T. ANDO, T. TANAKA AND M. KAMO

Deposition of diamond on nickel substrates has been performed by a microwave plasma reactor from methane–hydrogen gas mixtures. Growth features and the structures of the deposits have been studied as functions of methane concentration (0.3–5.0% (by volume)) and substrate temperature (700–1000 °C). At methane concentrations lower than 0.9%, diamond crystals, which have epitaxial relation to the substrate, have been observed to grow both on (111) and (100) faces of nickel. Other phenomena not observed with usual substrates have also been noted, and are believed to be caused by the unique properties of nickel.

1. Introduction

One of the characteristic features of the growth of diamond from the vapour phase is that diamond can be deposited in film forms on various non-diamond substrates. At present, however, what are known as diamond films are polycrystals. The necessity and advantages of single crystal films being obvious, attempts have been made to prepare single crystal films or epitaxial films on various non-diamond substrates.

Jeng & Tuan (1990) reported that clusters of oriented cubic nucleation of 1–2 μm and local epitaxial crystals of 120 × 150 μm on silicon {100} have been obtained by pretreating the substrate *in situ*, followed by usual growth process with a microwave plasma-assisted chemical vapour deposition. Koizumi *et al.* (1990) have reported epitaxial growth takes place on cubic boron nitride {111} surface by DC plasma chemical vapour deposition (CVD). Later, they found that epitaxy occurs on $\{111\}_B$ surface whose topmost layer is composed only of boron, while no tendency to epitaxy was observed with $\{111\}_N$ surface whose topmost layer consists entirely of nitrogen (Koizumi *et al.* 1991). Prins & Gaigher (1991) made an experimental study of ion implantation of carbon ions into copper and concluded that epitaxial diamond is formed at the surface by the precipitation of implanted carbon atoms under the influence of the substrate crystal structure. Stoner & Glass (1992) prepared what they termed textured diamond films on β–SiC by microwave plasma CVD preceded by *in situ* bias treatment that enhances nucleation. They found approximately 50% of the initial diamond nuclei are aligned with C(001) planes parallel to the SiC(001) plane.

This paper reports the results of further studies of the previous work (Sato *et al.*

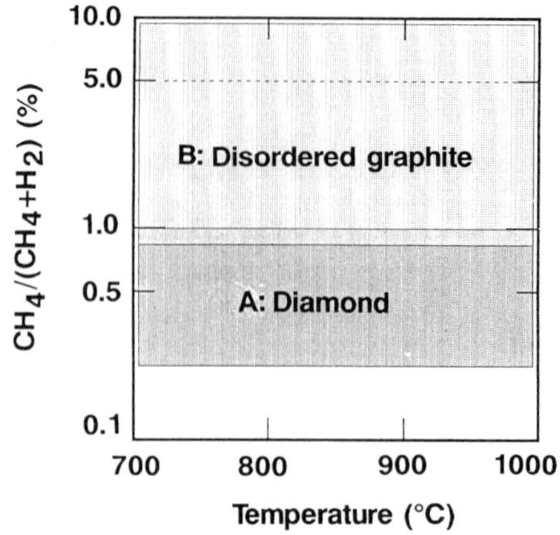

Figure 1. Schematic diagram of the relation between the deposition conditions and the structures of the deposits.

and cobalt substrates under appropriate gas compositions and substrate temperatures. Other phenomena characteristic to nickel substrates are also reported and discussed.

2. Experimental

A plasma reactor using microwave for excitation, similar to the one first adopted for diamond synthesis by Kamo *et al.* (1983), has been used for the growth studies. Briefly, it consists of a reaction chamber made of a quartz glass tube of 50 mm outer diameter, a gas feeding system, a simple mechanical pump for evacuation, and a microwave cavity connected to a microwave (2.45 GHz) generator. Gaseous mixtures of methane and hydrogen have been used for deposition. Nickel polycrystalline plates as well as (111) and (100) single crystals have been used as substrates. Deposition has been conducted under the following conditions: gas pressure within the range 40–100 Torr†; methane concentration within 0.2–5.0% (by volume); gas flow rate at 100 cm^3 min^{-1} (STP); and substrate temperature of 700–1000 °C.

The samples have been studied by optical microscopy, scanning electron microscopy (SEM), Raman spectroscopy and X-ray as well as electron diffraction. Raman scattering was measured with a Spex Ramalog 1403 with back scattering geometry. The 514.5 nm line of an argon ion laser was used for excitation.

3. Results

Structure of the deposits was found to be critically dependent on the gas composition as illustrated schematically in the diagram shown in figure 1, which was obtained at the gas pressure of 40 Torr. Similar diagrams as functions of the gas composition and substrate temperature (T_s) have been obtained in the range 40–100 Torr. Diamond crystals having epitaxial relation to the substrate have been observed to grow when the concentration of the methane gas was less than 0.9%

† 1 Torr ≈ 133 Pa.

Figure 2

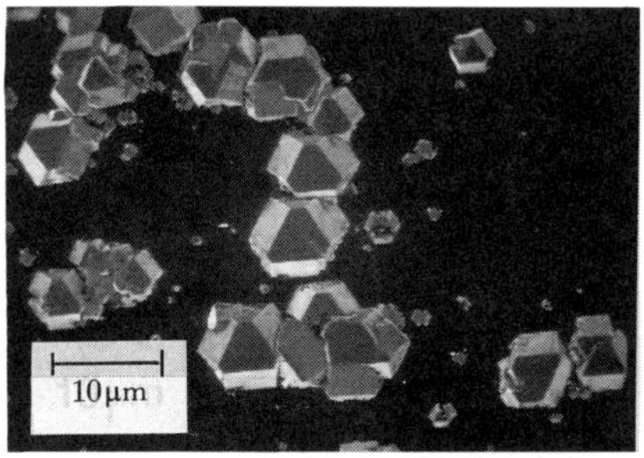

Figure 3

Figure 2. An SEM image of the epitaxial diamond crystals grown on (111) face of nickel grown at 0.5% methane, 100 Torr and substrate temperature $T_s = 880$ °C.

Figure 3. An SEM image of the epitaxial diamond crystals grown on (100) face of nickel grown at 0.5% methane, 60 Torr and at $T_s = 910$ °C.

Figure 4. An SEM image of coalesced epitaxial crystals on (111) face, grown at 0.5% methane, 100 Torr and $T_s = 880$ °C.

(region A in figure 1), and when the methane content was higher than 1.0% the initial deposits were found to consist entirely of disordered graphite, or soot-like carbons (region B in figure 1). In the region B at a methane concentration of 1.0–1.5%, separate diamond crystals started to grow after the substrate surface was covered with a layer of the soot-like carbons with thickness of 0.1–1.0 μm. No tendency to epitaxy has been observed with these crystals. Results similar to the above have also been observed with cobalt polycrystalline substrates.

(*a*) *Local epitaxy and coalescence of crystals*

For the samples prepared in region A of figure 1, individual diamond crystals were found to grow epitaxially on (111) face as well as on the (100) face, as shown in figures 2 and 3 respectively. X-ray measurements by Laue method showed that they are epitaxial with respect to the substrate plane within the experimental error of about 2°. Various observed orientations of crystals found on different grains of polycrystalline nickel plates suggest that diamond grows epitaxially on crystallographic planes other than (111) and (100) planes, but the determination of other planes has not been made. From SEM observation, it was noted that there are crystals which have no epitaxial relation to substrate. The probability of finding

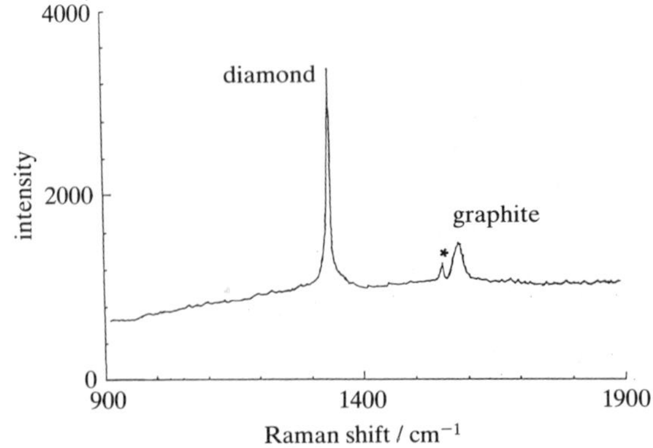

Figure 5. Raman spectrum of epitaxial diamond grown on (111) face. The smaller peak at 1580 cm^{-1} is due to well-crystallized graphite and a weak but sharp peak with asterisk is due to oxygen molecules in air.

these crystals decreases as the substrate temperature increases, and the samples prepared at a substrate temperature higher than about 880 °C are substantially free from non-epitaxial crystals.

As shown in figure 4, coalescence of epitaxial single crystals takes place where the nucleation density is high, leading to a larger single crystal area. The nucleation behaviour appeared to be similar to that of the substrates which are widely used, e.g. silicon, molybdenum, tantalum and tungsten. Surface treatments with hard abrasive powders, in particular diamond, were found to be effective in increasing the nucleation density. It has been noted that the nucleation density is lower for the samples deposited at higher substrate temperatures.

The Raman spectrum of an epitaxially grown crystal is shown in figure 5. The sharp, strong line at 1333 cm^{-1} is due to diamond (Solin & Ramdas 1970) and a weak peak at 1580 cm^{-1} is assigned to well-crystallized graphite (Tuinstra & Koenig 1970; Nemanich & Solin 1979). The line width observed with a number of crystals expressed as full width at half maximum (FWHM) was found to be within 1.8–2.0 cm^{-1}.

(*b*) *Disintegration of crystals*

A phenomenon that we have not encountered so far with other substrates of various types, including silicon, tungsten, tantalum, molybdenum, copper, silica, alumina, some carbides and nitrides, has been noted with nickel as well as cobalt. A crystal once grown epitaxially disintegrates into smaller fragments of epitaxial crystals which tend to disappear. This has been noted when deposition was interrupted for the inspection of the sample and then the deposition was resumed. After the second run, crystals that had grown in the first run were found to disintegrate as shown in figure 6. Similar crystals have been also found after a long deposition run of about 40 h or more.

(*c*) *Formation of well-crystallized graphite*

Formation of a thin but visible layer has been often noted at the back side of the substrate which faced the surface of the sample holder made of silica glass. It appeared that thicker layers were formed when the deposition was conducted at

Figure 6

Figure 7

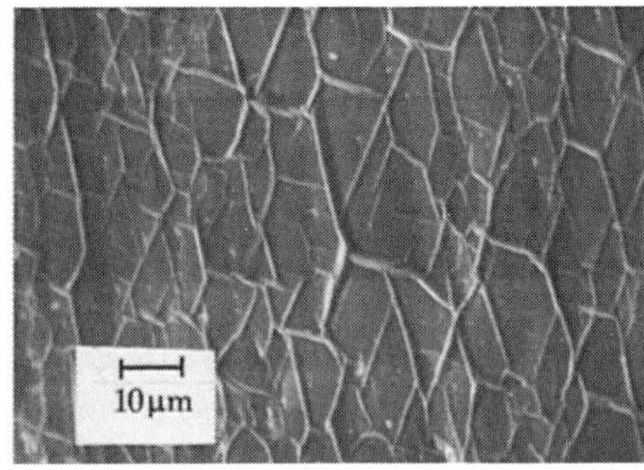

Figure 6. An SEM image of a partly disintegrated crystal.

Figure 7. An SEM image of well-crystallized graphite layer formed at the back side of nickel substrate which has been subjected to deposition at 0.5% methane, 100 Torr and $T_s = 980$ °C.

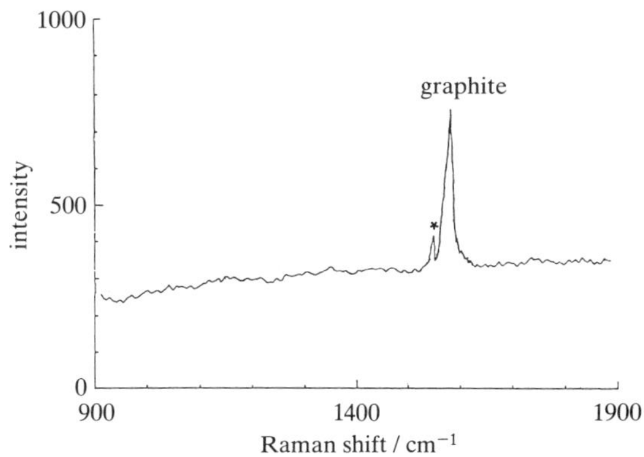

Figure 8. Raman spectrum of the graphite layer formed at the back side of the substrate under the same condition as that of the sample shown in figure 7.

higher temperatures. An SEM image is shown in figure 7 and the Raman spectrum of the layer in figure 8, which consists of a single sharp line at 1580 cm^{-1}, indicating that the layer is well-crystallized graphite. It has also been shown that the top surface of the substrate, where diamond deposits, is also covered with a very thin layer of good quality graphite which is normally invisibly thin but gives an appreciable Raman signal, as seen in figure 5. Raman microprobe measurements indicated that a thin graphite layer is also present at the interface of diamond and substrate.

4. Discussion

It has been shown that individual crystals having epitaxial relation to the substrate grow when the gas composition is controlled below the critical methane content of about 0.9%. Further, the possibility has been shown that the coalescence of these crystals can take place to form a single crystal film.

It has also been shown, however, that diamond crystals start to disintegrate into smaller crystals after interruption or after a prolonged reaction and this disturbs an extensive growth and coalescence of crystals. Here, a brief discussion will be given of the possible causes of anomalous phenomena observed with nickel and also of the future directions.

(*a*) *Causes of anomalous deposition features*

Apart from the epitaxial growth of individual crystals, anomalous reaction features have been noted. One is the fact that when the methane content is higher than about 1.0%, the initial deposits consist entirely of disordered graphite (region B of figure 1). The abrupt change in structure from diamond to graphite within a narrow range of about 0.1% is in sharp contrast to the gradual changes in the fraction of graphite content as a function of methane content (or carbon content, in general) in the gas phase observed with most of the well-known substrates (see, for example, Sato & Kamo 1989; Bachmann *et al.* 1991).

The other anomaly is the disintegration of diamond crystals. Though no definite conclusions have been drawn as to the causes of these anomalies, it seems certain that they are related to the unique properties of nickel, of which those relevant to the present work may be summarized as follows. In the temperature range of the present deposition experiments (700–1000 °C), nickel is known to show, (i) catalytic activity to carbon–hydrogen compounds, (ii) high solid solubility to carbon which increases with increasing temperature, (iii) high diffusion rate of carbon, (iv) very high diffusion rate of hydrogen and (v) little tendency to form stable carbides.

The formation of disordered graphite at a relatively low methane concentration of about 1.0% may be caused by enhanced rate of carbon deposition resulting from the catalytic activity of nickel. It may be plausible to think that the deposit on nickel has a structure corresponding to that of the deposit on silicon prepared from a mixed gas of much higher methane concentration due to the catalytic activity.

One of the probable mechanisms of disintegration of diamond crystals is that it is caused by the transport of carbon atoms from diamond to graphite through nickel, since the free energy of diamond is higher than that of graphite. As described above, well crystallized graphite is found at the back side of the substrate. The graphite can be formed on gradual cooling after the microwave power is turned off. Thus the subsequent deposition run with the same sample will certainly be subject to this mechanism. If graphite is also formed while deposition is conducted, it explains the degradation in prolonged deposition runs.

As immediate technical approaches to prevent the graphite formation, the following have been considered: (i) to expose the whole surface area of a substrate to the gas plasma; and (ii) to minimize temperature variation as functions of reaction time and position of the substrate. For extensive coalescence, increase of nucleation density and increase of growth rate will be effective.

In the above discussion, it is postulated that a thermodynamical equilibrium approach is valid inside the substrate. On the other hand, it may be obvious that kinetic principles predominate in the reactions occurring at the surface and in the gas phase near the substrate surface where a sharp gradient in temperature exists.

(*b*) *Advantages of nickel substrates*

It is possible to point out some advantages in using nickel as substrates even at this stage, where the technique still needs further developments. (i) It is feasible to obtain large area single crystal substrates by various ways that include growth of bulk crystals and epitaxial films. (ii) It is feasible to attain a close matching of the lattice parameters at a deposition temperature by alloying with other metals and elements, thus enabling to obtain epitaxial films substantially free from dislocations coming from the lattice parameter difference. (iii) Epitaxial diamond films will be readily separated from the substrates by intentional precipitation of graphite at the interface of nickel and diamond, thus allowing repeated usage of the bulk single crystal substrates, and ready transfer of the films onto other substrates for further growth and/or manipulation.

We believe that the epitaxy on nickel and its alloys is worth further study. Because of the seemingly complex reactions involved with nickel, as discussed above, it is planned to make experimental studies to elucidate some of the fundamental processes in the reactions and transport phenomena, together with technical approaches, also suggested above. The same may be concluded for cobalt.

References

Bachmann, P. K., Leers, D. & Lydtin, H. 1991 Towards a general concept of diamond chemical vapour deposition. *Diamond Related Mater.* **1**, 1.

Jeng, D. G. & Tuan, H. S. 1990 Oriented cubic nucleation and local epitaxy during diamond growth on silicon {100} substrates. *Appl. Phys. Lett.* **56**, 1968.

Kamo, M., Sato, Y., Matsumoto, S. & Setaka, N. 1983 Diamond synthesis from gas phase in microwave plasma. *J. Cryst. Growth* **62**, 642.

Koizumi, S., Murakami, T. & Inuzuka, T. 1990 Epitaxial growth of diamond thin films on cubic boron nitride {111} surfaces by dc plasma chemical vapor deposition. *Appl. Phys. Lett.* **57**, 563.

Koizumi, S., Murakami, T. & Inuzuka, T. 1991 An investigation of apitaxial growth of diamond on cBN surfaces. In *Proc. 5th Symp. on Diamond* (ed. M. Yoshikawa), p. 4. New Diamond Forum of Japan.

Nemanich, R. J. & Solin, S. A. 1979 First- and second-order Raman scattering from finite-size crystals of graphite. *Phys. Rev.* B **20**, 392.

Prins, J. F. & Gaigher, H. L. 1991 A TEM study of layers grown on copper using carbon-ion-implantation. In *Proc. 2nd Int. Conf. on New Diamond Science and Technology* (ed. R. Messier, J. T. Glass, J. E. Butler & R. Roy), p. 561. Pittsburgh: Materials Research Society.

Sato, Y. & Kamo, M. 1989 Texture and some properties of vapor deposited diamond. *Surf. Coatings Technol.* **40**, 183.

Sato, Y., Yashima, I., Fujita, H., Ando, T. & Kamo, M. 1991 Epitaxial growth of diamond from the gas phase. In *Proc. 2nd Int. Conf. New Diamond Science and Technology* (ed. R. Messier, J. T. Glass, J. E. Butler & R. Roy), p. 371. Pittsburgh: Materials Research Society.

Solin, S. A. & Ramdas, A. K. 1970 Raman spectrum of diamond. *Phys. Rev.* B **1**, 1687.

Stone, B. R. & Glass, J. T. 1992 Textured diamond growth on (100) β–SiC via microwave chemical vapor deposition. *Appl. Phys. Lett.* **60**, 698.

Tuinstra, F. & Koenig, J. L. 1970 Raman spectrum of graphite. *J. Chem. Phys.* **53**, 1126.

Comment

M. Seal (*Sigillum B.V., Amsterdam, The Netherlands*). A possible way of avoiding the loss of the epitaxial diamond film through dissolution in the nickel is the following. After formation of the initial oriented diamond layer, stop the diamond growth and deposit a different support film (e.g. silicon) over the growth surface. Dissolve or release the nickel, and remove the graphite and any residual nickel from the lower surface of the diamond by chemical or ion etching. One might then be left with an oriented diamond layer on a silicon substrate, which could be thickened, hopefully epitaxially, by any of the standard diamond CVD techniques.

5

The optical and electronic properties of semiconducting diamond

BY ALAN T. COLLINS

In this paper I review the evidence that shows that the optical and electronic properties of semiconducting diamond can be understood in terms of boron acceptors partially compensated by deep donors. In natural semiconducting diamond, in which the total impurity concentration is less than 1 ppm, there is a lot of fine structure in the acceptor absorption spectrum that is not fully understood, and the electrical conductivity is primarily associated with the thermally activated excitation of holes from the acceptor ground state to the valence band. Some of the problems regarding the analysis of Hall effect data in this material are discussed, including the temperature dependences of the scattering mechanisms, of the contribution from the split-off valence band and of the population of excited states. There are no adequate theoretical descriptions of any of these processes, and this leads to some uncertainties in the values of the parameters derived from the temperature dependence of the Hall coefficient. For boron-doped synthetic diamond, and thin film diamond grown by chemical vapour deposition (CVD), the defect concentrations are generally much higher, and much more inhomogeneous, than in natural semiconducting diamond. This results in a substantial broadening of the acceptor absorption spectrum and the electronic properties are greatly modified by increasing contributions from impurity band conduction as the acceptor concentrations are increased, leading to very low mobility values. For both polycrystalline and single crystal homoepitaxial CVD diamond, measurements of the electrical properties can be completely invalidated by the presence of a surface layer of non-diamond carbon.

1. Introduction

Since the speculation by Davis *et al.* (1988) and others, that thin film semiconducting diamond exhibits considerable potential for opto-electronic devices, there has been renewed intensive activity in the study of the optical and electronic properties of diamond, following almost two decades of relative quiescence. In this review I consider those aspects of the early work that have particular relevance to contemporary studies and briefly examine some of the conclusions of recent investigations.

Natural semiconducting diamond (type IIb diamond) was first reported by Custers (1952). Electrical transport measurements (for a review see Collins & Lightowlers 1979) established that this material is a partially compensated semiconductor, and that the acceptor is boron with an ionization energy of 0.368 eV. That work is discussed further in §3*a*. Nitrogen acts as a deep donor in diamond, and is normally the major impurity in both natural and synthetic diamond; these diamonds are therefore electrical insulators. (I use the word synthetic to mean those

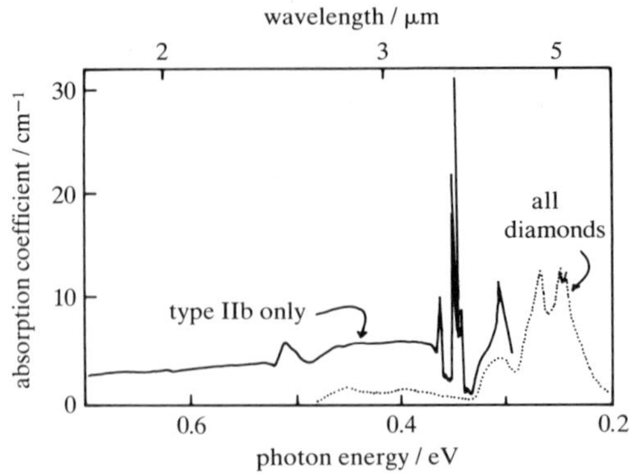

Figure 1. Acceptor absorption in a natural type IIb diamond at 77 K. Redrawn from Smith & Taylor (1962).

diamonds grown by the high temperature, high pressure process.) Huggins & Cannon (1962) first showed that semiconducting type IIb diamond can be grown by excluding nitrogen and incorporating boron during growth. This material had the morphology and dimensions of abrasive grit and presented a considerable challenge in the determination of its optical properties (Collins *et al.* 1965) and electronic properties (Lightowlers & Collins 1966; Williams *et al.* 1970).

2. Optical properties

Luminescence spectra from diamond, including bound exciton recombination and the 'band A' system in natural and synthetic semiconducting diamond, have been reviewed recently by me (Collins 1992). In this section I therefore consider only the optical absorption associated with the boron acceptor.

(*a*) *Natural type IIb diamond*

Transitions from the boron acceptor to the valence band give rise to a series of excited state transitions which start at 0.304 eV and merge with the photoionization continuum at about 0.37 eV. Figure 1 shows this additional absorption produced in a type IIb diamond (Smith & Taylor 1962), superimposed on the two- and three-phonon combination bands which are present in all diamonds. The photoionization continuum extends into the red part of the visible spectrum, and gives type IIb diamonds their characteristic blue colour. Absorption due to the higher-energy excited states is shown in more detail in figure 2. Six of the peaks in this spectrum disappear when the sample is cooled to 4.2 K and Crowther *et al.* (1967) have associated this effect with a 2.1 meV splitting of the acceptor ground state. These latter authors carried out uniaxial stress measurements on the acceptor spectrum, and were able to propose a plausible classification scheme for the peaks at energies below 0.348 eV, but the full complexity of this spectrum is still not understood.

(*b*) *Synthetic type IIb diamond*

Absorption spectra made on synthetic semiconducting diamond by Collins *et al.* (1965) resembled the spectrum shown in figure 1, but even at low neutral acceptor concentrations the peaks were much broader and the fine structure shown in figure

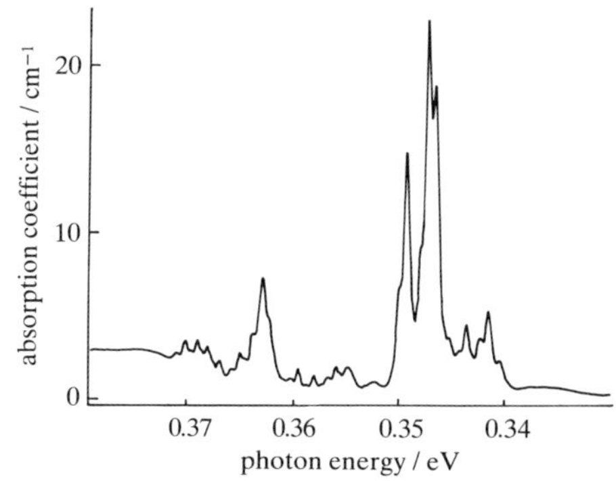

Figure 2. Excited state region of the acceptor spectrum in a natural type IIb diamond at 77 K. Redrawn from Lawson (1991).

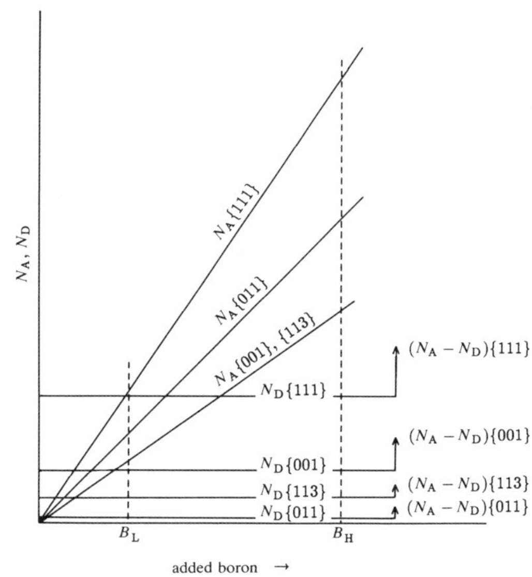

Figure 3. Relation between the nitrogen donor concentration, the boron acceptor concentration and the neutral acceptor concentration in the major growth sectors of synthetic diamond, shown as a function of the added boron concentration. See text for details. Redrawn from Burns *et al.* (1990).

2 could not be detected. As the acceptor concentration increased (assessed semi-quantitatively from the depth of blue coloration association with the photoionization continuum) the peaks became progressively smeared out and merged with the photoionization continuum. Type IIb diamonds with much better quality than those used by Collins *et al.* may now be grown by the temperature gradient method (Strong & Chrenko 1971), but in most cases the absorption peaks are still very much broader than those observed in natural type IIb diamond; furthermore the broadening of the spectrum is far more severe in some growth sectors than others (Burns *et al.* 1990).

To understand this latter phenomenon we need to consider the way different impurities are incorporated into the different growth sectors of synthetic diamond. This can be summarized on a diagram of the type originally proposed by Kanda *et al.* (1987) and extended by Burns *et al.* (1990), shown in figure 3. This diagram will also have considerable implications for the boron doping of diamond grown by chemical vapour deposition (CVD).

The four horizontal lines on figure 3 indicate the relative concentrations of isolated substitutional nitrogen in the major growth sectors; the {111} sectors contain most nitrogen, and the {011} sectors least. Optical transitions from the nitrogen donor to the conduction band give synthetic diamonds a characteristic yellow colour, and in a polished section of diamond the different nitrogen concentrations in the different growth sectors are evident from the depth of yellow coloration. The gradients of the diagonal lines on figure 3 show the rates at which boron is incorporated into the different growth sectors. A given growth sector becomes semiconducting when the boron concentration exceeds the nitrogen concentration. The horizontal lines are effectively the origins for the *uncompensated* acceptor concentrations ($N_A - N_D$) in the corresponding growth sectors, as indicated on the right of the diagram. We see that the {011} sector is the first to acquire semiconducting properties, and that at relatively low boron concentrations, B_L, ($N_A - N_D$) is highest in this growth sector. Again this effect is visible by eye, because at boron concentrations between 0 and B_L some growth sectors are blue while others are still yellow. At higher boron concentrations, B_H, the neutral acceptor concentration is highest in the {111} sectors.

If there is a significant source of nitrogen in the growth capsule, and a sufficient quantity of boron is added to produce semiconducting diamond, then, although the electrically active boron concentration may be small, the total impurity concentration can be quite large. The relatively high concentration of defects, compared with less than 1 ppm in natural type IIb diamond (see §3*a*) almost certainly accounts for the much greater linewidths observed in the acceptor absorption spectrum (Rooney 1992). This effect is expected to be most pronounced in the {111} growth sectors (see figure 3), as observed experimentally (Burns *et al.* 1990; Rooney 1992). If the nitrogen in the growth capsule is very efficiently 'gettered', by adding materials such as Ti and Zr, the resulting crystals are sometimes semiconducting by virtue of the 'accidental' boron present; in one such crystal grown from ^{13}C, examined by Lawson (1991), the linewidths in the acceptor spectrum were very much sharper than those in the specimens examined by Burns *et al.* (1990), and it was possible to determine the isotope shifts of the major features in the spectrum.

(*c*) *Semiconducting* CVD *diamond*

For CVD diamond grown in a low pressure plasma the nitrogen content is normally very low, but can, nevertheless, be detected using luminescence spectroscopy (Collins 1992). Semiconducting CVD diamond may readily be produced by doping with boron in one of the ways discussed later in §3*d*. The experience with synthetic diamond described in §2*b* suggests that the boron distribution in polycrystalline CVD diamond (which certainly contains both {111} and {001} growth sectors) will be inhomogeneous, and that the rate of incorporation of boron in homoepitaxial material will be greatly influenced by the orientation of the substrate. We shall see below that this is indeed the case.

3. Electrical transport measurements

Most of our information about the electronic properties of semiconducting diamond has come from Hall effect and conductivity studies (for reviews see Kemmey & Wedepohl 1965; Collins & Lightowlers 1979). Here I re-iterate and extend some of this earlier work.

(*a*) *Hall effect measurements*

Hall effect measurements on natural semiconducting diamond have generally been made with the sample geometry in its traditional form of two contacts on opposite faces of a rectangular block. After a correction has been made for the finite length to width ratio of the specimen (Isenberg *et al.* 1948) the concentration p of free holes may be determined from the Hall coefficient R using

$$p = r/Re, \tag{1}$$

where r is the ratio of the Hall mobility μ_H to the conductivity mobility μ_c and e is the electronic charge.

The value of r depends on the scattering processes; for a non-degenerate semiconductor a number of simplifying assumptions lead to $r = \frac{3}{8}\pi = 1.18$ for acoustic phonon scattering and $\frac{315}{512}\pi = 1.93$ for ionized impurity scattering (Sze 1981). Phonon scattering is the dominant mechanism at high temperatures and analysis suggests the mobility will vary as $T^{-\frac{3}{2}}$, compared with $T^{\frac{3}{2}}$ for ionized impurity scattering (Sze 1981). In practice the high-temperature mobility of Si, Ge and diamond varies as T^{-S} with S around 2.6 for Si and 2.8 for diamond. This brief summary indicates one (of many) problems in fitting Hall effect data to theoretical formulae. The parameter r is almost certainly temperature dependent, and there is no obvious way to calculate the actual value. Generally r is set to 1 or $\frac{3}{8}\pi$ in the analysis of the data, and variations between the calculated and experimental data may be attributed at least in part to the temperature dependence of r.

For a partially compensated p-type semiconductor, with an acceptor concentration sufficiently small that there is no degeneracy, the hole concentration p at temperature T may be written as (Blakemore 1962)

$$\frac{p(p+N_D)}{(N_A-N_D-p)} = \left(\frac{2}{g_a}\right)\left(\frac{2\pi m^* kT}{h^2}\right)^{\frac{3}{2}} \exp\left(-E_A/kT\right), \tag{2}$$

where N_A and N_D are the acceptor and donor concentrations, g_a is the ground state degeneracy factor for the acceptor, m^* is the density of states effective mass for the holes and E_A is the acceptor ionization energy.

Mitchell (1963) proposed that g_a should be set to 2; however, setting $g_a = 2$ implies that there is a single, spherically symmetrical valence band with a scalar effective mass m^*. For acceptors in silicon and germanium, Sze (1981) states that g_a should be 4 because the valence band is doubly degenerate at $k = 0$. In diamond the situation is more complicated; Rauch (1962) has shown from cyclotron resonance studies that two bands are degenerate at $k = 0$ with masses $0.70 \pm 0.01\ m_e$ and about $2.1\ m_e$, with a third band split off by 6 ± 1 meV with a mass of $1.06\ m_e$, where m_e is the mass of a free electron. Thus, in equation (2) we should certainly set $g_a = 4$, and at high temperatures $g_a = 6$ may be more appropriate. Assuming a simple Boltzmann distribution, the contribution from the split-off band will vary from 60 to 95% in the temperature range 150–1250 K covered in figure 4.

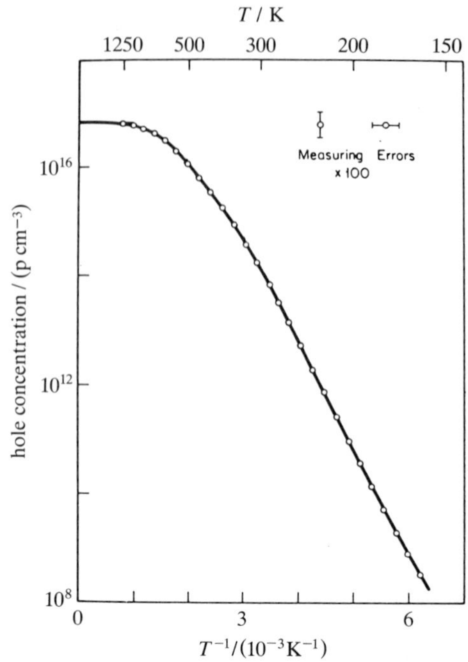

Figure 4. Hole concentration in a natural semiconducting diamond as a function of reciprocal temperature, derived from the Hall coefficient (equation (1)) with $r = \frac{3}{8}\pi$. Redrawn from Collins & Lightowlers (1979).

Note from equation (2) that when $T \to \infty$, p tends to $N_A - N_D$. In practice, for the typical acceptor and donor concentrations of 5×10^{16} and 5×10^{15} cm^{-3} encountered in natural diamond, p is within 2 or 3% of saturation at $T = 1250$ K. This temperature represents a realistic upper limit for Hall effect measurements on diamond, and for the more heavily doped synthetic and CVD specimens it becomes progressively more difficult to approach the saturated carrier concentration as the acceptor concentration is increased.

At low temperatures $p \ll N_A$ or N_D, and, provided the acceptor concentration is sufficiently low to avoid impurity band conduction (discussed later), equation (2) approximates to

$$p \propto T^{\frac{3}{2}} \exp(-E_A/kT). \tag{3}$$

From measurements on 5 diamonds Collins & Williams (1971) quote an average value of $E_A = 368.5 \pm 1.5$ meV, determined using (3) in the temperature interval 160–330 K.

At intermediate temperatures there is a further complication that may make it difficult to fit the data using (2). An acceptor impurity is neutral whether it has a hole bound in the ground state at energy E_A above the valence band, or in an excited state rather closer to the valence band. This situation is discussed by Blakemore (1962). In diamond, because there is a superficial similarity of the energies of the excited states with those calculated by the Bohr model, it may be necessary to take the

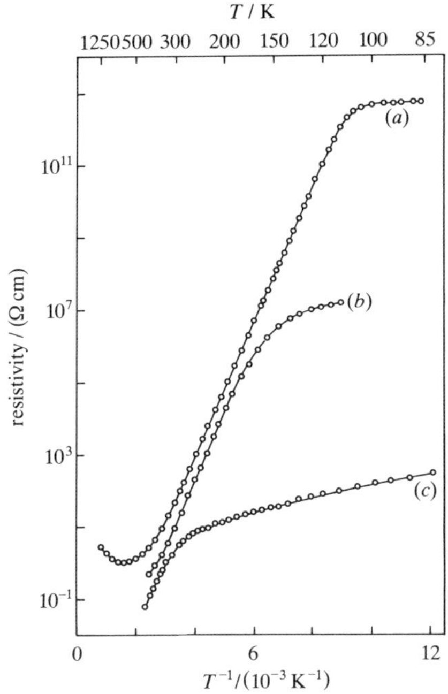

Figure 5. Resistivity as a function of reciprocal temperature for (*a*) a natural type IIb diamond, (*b*) a synthetic type IIb diamond with $(N_A - N_D) = 3 \times 10^{17}$ cm^{-3} and (*c*) as *b* with $(N_A - N_D) = 10^{18}$ cm^{-3}. Redrawn from Collins & Lightowlers (1979).

population of these states into account when analysing the temperature dependence of the carrier concentration.

Following Mitchell (1963), Collins & Williams (1971) disregarded all the complexities described above and based their analyses on (2) with $g_a = 2$. They found that when the values for N_A, N_D and E_A that gave the best fits were substituted into (2), and m^* calculated from the experimental data over the range of temperatures used, this effective mass appeared to be constant at around $0.75\, m_e$ below room temperature, but decreased appreciably at higher temperatures in a non-monotonic fashion. In the light of the above discussion we recognize that m^* has simply been used as an adjustable parameter, and that much of the apparent variation with temperature is due to the temperature variation of r, to the temperature dependence of the contribution from the split-off band, to the neglect of the contribution from excited states and to a probable small variation of E_A with temperature.

(*b*) *Resistivity measurements*

In figure 5 the resistivity is shown as a function of reciprocal temperature for a natural diamond and two synthetic diamonds. Dealing first with the natural diamond (*a*) we see that at around 650 K there is a shallow minimum in the resistivity curve; this is because the carrier concentration has almost reached saturation (figure 4) but the mobility is still decreasing at higher temperatures. Between about 400 K and 120 K the log (resistivity) increases linearly with an activation energy of 0.37 eV, and at 120 K there is a sharp knee beyond which the resistivity increases very gradually at lower temperatures. Curve (*b*) for the first synthetic diamond has a substantial linear section with a change of slope to a lower value at about 160 K, and curve (*c*) for the second synthetic diamond has a short section, parallel to that for the other two diamonds, and a change of slope to a lower value at about 280 K. The neutral acceptor concentrations for (*b*) and (*c*) were estimated optically to be 3×10^{17} cm^{-3} and 1×10^{18} cm^{-3} respectively (Williams *et al.* 1970). These latter authors based their analysis of these data on the model discussed earlier by Davis & Compton (1965). The conductivity may be expressed as a sum of three terms

$$\sigma = \sigma_1 \exp(-E_1/kT) + \sigma_2 \exp(-E_2/kT) + \sigma_3 \exp(-E_3/kT). \tag{4}$$

The activation energy E_1 is the normal acceptor ionization energy, associated with transitions from the acceptor ground state to the valence band, and is observed in all samples provided the acceptor concentration is not too high. The activation energy E_2 can only be observed in the intermediate concentration range and is associated with conduction in an impurity band. When the acceptor concentration is small E_2 is close to E_1, but when the acceptor concentration is increased so that there is an appreciable overlap between the wavefunctions of neighbouring acceptor centres E_2 is reduced. Finally, at the acceptor concentration for which the metal–insulator transition occurs, $E_2 \rightarrow 0$. The activation energy E_3 is most prominent for specimens with a relatively low impurity concentration, and is interpreted in terms of the energy associated with the tunnelling transition of a hole from an unoccupied to an occupied acceptor site.

In figure 5*a* the change of slope in the resistivity curve for the natural diamond is interpreted as tunnelling conductivity with activation energy E_3 (Collins & Williams 1971), whereas all three conductivity mechanisms are believed to be operative for the synthetic diamonds in figure 5*b*, *c* (Williams *et al.* 1970). The latter authors were also able to demonstrate that the Hall coefficient for their synthetic diamonds went through a maximum at the onset of impurity band conduction, as expected theoretically (Davis & Compton 1965).

Although the work of Williams *et al.* appears plausible it is now recognized that the distribution of acceptor and donor impurities in synthetic diamond is very dependent on the growth sector (§2*b*), and it would be dangerous to try to extract too much information from these early measurements.

(*c*) *Differential capacitance measurements*

If the capacitance C of a Schottky barrier diode is measured as a function of the reverse voltage V for a p-type semiconductor, the uncompensated acceptor concentration is given by

$$N_A - N_D = 2[A^2 \epsilon e \, \mathrm{d}(1/C^2)/\mathrm{d}V]^{-1}, \tag{5}$$

where A is the area of the diode, ϵ is the permittivity and e the electronic charge. This method has been used to measure the acceptor concentrations in both synthetic (Glover 1973) and natural (Lightowlers & Collins 1976) semiconducting diamond. It is also an ideal method to study thin films of CVD diamond, since the depletion layer thickness in a reverse-biased diode is less than 1 μm for acceptor concentrations around 5×10^{16} cm^{-3}. In the natural diamonds examined by Lightowlers & Collins the ratios N_D/N_A determined from Hall effect measurements lay in the range 3 to 50.

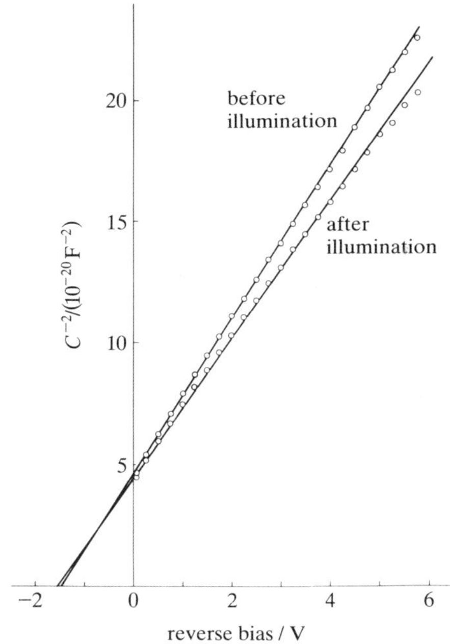

Figure 6. $1/C^2$ against V plots for a 0.5 mm diameter gold Schottky barrier diode on a natural semiconducting diamond before and after illumination which neutralizes the compensated donors. The lines are least squares fits to the experimental points. From Lightowlers & Collins (1976).

These authors showed that by illuminating a reverse biased Schottky diode with visible light the compensated donors could be neutralized. Hence a plot of $1/C^2$ against V before and after illumination yields both $(N_A - N_D)$ and N_A respectively. Typical plots are shown in figure 6, and the ratio of N_D/N_A is comparable with that determined from the temperature dependence of the Hall effect. Lightowlers & Collins showed that the boron concentrations measured by a nuclear technique, in the same area as that probed by the Schottky diode, exhibited a good correlation with N_A in the five diamonds studied earlier by Collins & Williams (1971). The values of N_A were, however, almost a factor of 2 lower than those determined from the Hall effect measurements. This discrepancy may be associated with the oversimplified theoretical treatment of the Hall data, discussed earlier. However, to yield saturation values of the hole concentration almost a factor of 2 lower than those shown in figure 4 apparently requires reducing the high-temperature value of r in (1) to 0.6. This is outside the range calculated from standard scattering theories, but it must be remembered that these theories also fail to predict the correct value of S in the T^{-S} temperature dependence of the mobility.

(*d*) *CVD diamond*

Semiconducting CVD diamond may be made very easily by incorporating boron during growth. Many research groups have produced polycrystalline material, usually on silicon substrates, that have enabled crude rectifiers to be constructed (see, for example, Landstrass *et al.* 1991). However, the electrical transport properties of these films are difficult to interpret because of the scattering of carriers at grain boundaries, and because the non-diamond carbon frequently present at the grain boundaries may be electrically conducting (Shiomi *et al.* 1990; Muto *et al.* 1991). We will restrict our attention, therefore, to thin films of boron doped diamond grown epitaxially on a diamond substrate.

In many cases the techniques used for doping are rudimentary; for example, Gildenblat *et al.* (1990) either place boron powder on the substrate holder, or periodically insert a boron rod into the microwave plasma. Janssen *et al.* (1992) produced semiconducting diamond inadvertently as the plasma in their hot filament reactor slowly decomposed the hexagonal BN substrate holder. On the other hand Shiomi *et al.* (1991) have adjusted the B concentration in their samples in a more rigorous manner by adding controlled amounts of B_2H_6 to the gas stream.

In view of the experience with high-pressure synthetic diamond (§2*b*) it is not surprising that the growth rate of epitaxial CVD diamond, and the degree of boron incorporation, depend on the orientation of the substrate (Janssen *et al.* 1992). Cathodoluminescence measurements show that the boron doped films are frequently inhomogeneous (Janssen *et al.*); furthermore the photoluminescence band in the Raman spectrum of a boron doped layer grown by Grot *et al.* (1991) shows clear evidence that this material contains optical centres involving nitrogen and vacancies (Collins 1992). Thus, although it is possible to grow boron doped single crystal diamond by the CVD process, the quality of the films is not comparable with that of the carefully selected natural type IIb diamonds studied by Collins & Williams (1971) and Lightowlers & Collins (1976).

Gildenblat *et al.* (1990) made the important observation that when their homoepitaxial diamond layers were removed from the microwave reactor there was little difference between the sheet resistance of undoped and boron doped samples. However, after chemical cleaning in an oxidizing solution the sheet resistivity of the undoped layers was greater than 10^8 kΩ per square, while that of the boron doped layers remained low at 60 kΩ per square or less. Gildenblat *et al.* attributed this behaviour to the presence of a surface conducting layer. If, after chemical cleaning, the high resistivity undoped layers were re-introduced to the hydrogen plasma in the microwave reactor their resistivity again dropped to a low value. This reduction in resistivity was also observed when an insulating natural type IIa diamond was exposed to the hydrogen plasma. Albin & Watkins (1990) have proposed that the increased conductivity on hydrogenation is due to the diffusion of hydrogen into the diamond and the passivation of deep traps. The simpler explanation of a surface conducting layer that can be removed by chemical cleaning (Gildenblat *et al.*) carries more conviction.

Gildenblat *et al.* have used the differential capacitance technique described above to measure acceptor concentrations in their films, and show a $1/C^2$ against V plot which yielded an acceptor concentration of 3.2×10^{17} cm^{-3}. Temperature dependence measurements on the series resistance of another Schottky diode yielded an activation energy of 0.15 eV in the temperature range 26–580 °C. The same research group has used a method of selective deposition to grow a film with the conventional Hall effect geometry (Grot *et al.* 1991). Here the temperature dependence of resistivity yielded an activation energy of 0.31 eV, and room-temperature Hall effect data indicated a hole concentration of 10^{14} cm^{-3} and a Hall mobility of 290 cm^2 V^{-1} s^{-1}. This compares with values of 1200 to 2000 cm^2 V^{-1} s^{-1} obtained on high-quality natural type IIb diamond (Collins & Lightowlers 1979).

Other groups (for example Geis 1990; Shiomi *et al.* 1991) have also made electrical transport measurements on boron doped homoepitaxial CVD diamond and found acceptor activation energies ranging from 0.002 to 0.39 eV, and mobilities which are much lower than those obtained with natural type IIb diamond.

Much of the behaviour observed for boron doped CVD diamond is consistent with the analysis of impurity band conduction in synthetic diamond by Williams *et al.* (1970). However, in view of the findings of Gildenblat *et al.* of a conducting surface layer present on as-grown films, the relatively poor quality of many films (as evidenced from photoluminescence and cathodoluminescence measurements (Collins 1992; Janssen *et al.* 1992)) and the known experimental problems with making reliable electrical transport measurements on diamond, much of the information in the literature needs to be approached with a considerable degree of caution.

4. Conclusions

All of the experimental work on natural, synthetic and CVD semiconducting diamond indicates that the optical and electronic properties may be understood in terms of boron acceptors, with an ionization energy of 0.368 eV, partially compensated by a smaller concentration of deep donors. The donors are isolated substitutional nitrogen atoms in synthetic diamond, and assumed to be nitrogen in natural diamond. The total impurity concentration in natural type IIb diamond is less than 1 ppm, and the much higher defect concentrations generally present in doped synthetic and CVD diamond lead to considerable broadening of the acceptor infrared absorption spectrum, to impurity band conduction and relatively low mobility values.

At a more detailed level the fine structure in the acceptor absorption spectrum is not properly understood. Furthermore, Hall effect data can only be analysed approximately because the temperature dependence of the scattering mechanisms, of the contribution from the split-off valence band and of the population of excited states cannot be determined with any certainty.

The distribution of neutral acceptors in synthetic and CVD diamond is extremely inhomogeneous, and it is important that the limitations of the samples, and of the theoretical understanding of the defect centres, are recognized in the ongoing research in this area.

References

Albin, S. & Watkins, L. 1990 In *Diamond, silicon carbide and related wide bandgap semiconductors* (ed. J. T. Glass, R. Messier & N. Fujimori). Materials Research Society Symposium Proceedings, vol. 162, pp. 303–308.

Blakemore, J. S. 1962 *Semiconductor statistics*. Oxford: Pergamon Press.

Burns, R. C., Cvetkovic, V., Dodge, C. N., Evans, D. J. F., Rooney, M-L. T., Spear, P. M. & Welbourn, C. M. 1990 *J. Crystal Growth* **104**, 257–279.

Collins, A. T. 1992 *Diamond Related Mater.* **1**, 457–469.

Collins, A. T., Dean, P. J., Lightowlers, E. C. & Sherman, W. F. 1965 *Phys. Rev.* A **140**, 1272–1274.

Collins, A. T. & Lightowlers, E. C. 1979 In *The properties of diamond* (ed. J. E. Field), pp. 79–105. London: Academic Press.

Collins, A. T. & Williams, A. W. S. 1971 *J. Phys.* C **4**, 1789–1800.

Crowther, P. A., Dean, P. J. & Sherman, W. F. 1967 *Phys. Rev.* **154**, 772–785.

Custers, J. F. H. 1952 *Physica* **18**, 489–496.

Davis, R. F., Sitar, Z., Williams, B. E., Kong, H. S., Kim, H. J., Palmour, J. W., Edmond, J. A., Ryu, J., Glass, J. T. & Carter, Jr, C. H. 1988 *Mater. Sci. Engng* B**1**, 77–104.

Davis, E. A. & Compton, W. D. 1965 *Phys. Rev.* **140**, A2183–A2194.

Geis, M. W. 1990 In *Diamond, silicon carbide and related wide bandgap semiconductors* (ed. J. T. Glass, R. Messier & N. Fujimori). Materials Research Society Symposium Proceedings, vol. 162, pp. 15–22.

Gildenblat, G. Sh., Grot, S. A., Hatfield, C. W., Wronski, C. R., Badzian, A. R., Badzian, T. & Messier, R. 1990 In *Diamond, silicon carbide and related wide bandgap semiconductors* (ed. J. T. Glass, R. Messier & N. Fujimori) Materials Research Society Symposium Proceedings, vol. 162, pp. 297–302.

Glover, G. H. 1973 *Solid State Electron.* **16**, 973–983.

Grot, S. A., Hatfield, C. W., Gildenblat, G. Sh., Badzian, A. R. & Badzian, T. 1991 In *New diamond science and technology* (ed. R. Messier, J. T. Glass, J. E. Butler & R. Roy), pp. 949–955. Pittsburgh: Materials Research Society.

Huggins, C. M. & Cannon, P. 1962 *Nature, Lond.* **194**, 829–830.

Isenberg, I., Russel, B. R. & Greene, R. F. 1948 *Rev. Sci. Instrum.* **19**, 685–688.

Janssen, G., van Enckevort, W. J. P., Vollenberg, W. & Giling, L. J. 1992 *Diamond Related Mater.* **1**, 789–800.

Kanda, H., Ohsawa, T. & Fukunaga, O. 1987 In *Abstracts of the Second Meeting of 'Diamond'*, pp. 23–24. Tokyo, Japan. (In Japanese.)

Kemmey, P. J. & Wedepohl, P. T. 1965 In *Physical properties of diamond* (ed. R. Berman), ch. 12. Oxford: Clarendon Press.

Landstrass, M. I., Moyer, D., Yokota, S. & Plano, M. A. 1991 In *New diamond science and technology* (ed. R. Messier, J. T. Glass, J. E. Butler & R. Roy), pp. 969–974. Pittsburgh: Materials Research Society.

Lawson, S. C. 1991 PhD thesis, University of London, U.K.

Lightowlers, E. C. & Collins, A. T. 1966 *Phys. Rev.* **151**, 685–688.

Lightowlers, E. C. & Collins, A. T. 1976 *J. Phys.* D **9**, 951–963.

Mitchell, E. W. J. 1963 In *Proc. First Int. Cong. on Diamonds in Industry* (ed. P. Greene), pp. 241–251. London: Industrial Diamond Information Bureau.

Muto, Y., Sugino, T. & Shirafuji, J. 1991 *Appl. Phys. Lett.* **59**, 843–845.

Rauch, C. J. 1962 In *Proc. Int. Conf. on Physics of Semiconductors* (ed. A. C. Stickland), pp. 276–280. London: The Institute of Physics and the Physical Society.

Rooney, M-L. T. 1992 *J. Crystal Growth* **116**, 15–21.

Shiomi, H., Nishibayashi, Y. & Fujimori, N. 1991 In *New diamond science and technology* (ed. R. Messier, J. T. Glass, J. E. Butler & R. Roy), pp. 975–980. Pittsburgh: Materials Research Society.

Shiomi, H., Tanabe, K., Nishibayashi, Y. & Fujimori, N. 1990 *Jap. J. appl. Phys.* **29**, 34–40.

Smith, S. D. & Taylor, W. 1962 *Proc. phys. Soc.* **79**, 1142–1143.

Strong, H. M. & Chrenko, R. M. 1971 *J. phys. Chem.* **78**, 1835–1843.

Sze, S. M. 1981 *Physics of semiconductor devices*, 2nd edn. New York: Wiley.

Williams, A. W. S., Lightowlers, E. C. & Collins, A. T. 1970 *J. Phys.* C **3**, 1727–1735.

6

The thermal conductivity of CVD diamond films

BY T. R. ANTHONY

The thermal conductivity of chemical vapour deposition diamond films is controlled by the microstructure, impurity content and carbon double bonds in the films. In high conductivity films, dislocation scattering is dominant at low temperatures, while phonon–phonon scattering limits the conductivity at room temperature. In lower quality films, hydrogen and metal impurities as well as carbon double bonds constrain the conductivity up to room temperature. Significant anisotropies and gradients in the thermal conductivity exist in some films because of their micro structure.

1. Introduction

In the early 20th century, experimental measurements showed that diamond has a high thermal conductivity at both room and liquid nitrogen temperatures. At room temperature, diamond conducts heat better than silver which has the highest thermal conductivity of any metal. Recently, polycrystalline diamond films have been grown by a chemical vapour deposition (CVD) process, that have remarkably high thermal conductivities up to 85% of those of single-crystal diamond (Morelli *et al.* 1991). Elementary theory shows that the thermal conductivity K of diamond is given by,

$$K = \tfrac{1}{3}C_v \Lambda V, \tag{1}$$

where C_v is the heat capacity of diamond at constant volume and Λ is the mean free path and V is the velocity of lattice vibrations (phonons) in diamond. The phonon velocity in diamond is higher than any other solid and generates the exceptional thermal conductivity of diamond. Equation (1) can be written in terms of the phonon scattering rate S which is the phonon velocity V divided by the phonon mean free path Λ to give

$$K = \tfrac{1}{3}C_v V^2/S = \beta/S, \tag{2}$$

where β is a constant that is independent of the diamond type since both the heat capacity and sound velocity vary negligibly with the diamond micro structure or isotopic composition (Anthony *et al.* 1992; Morelli *et al.* 1991).

Phonons are scattered in diamond by many different obstacles including vacancies, isotopes, impurities, double and triple carbon bonds (perfect diamond contains only single carbon bonds) dislocations, precipitates, stacking faults, grain boundaries, surfaces and other phonons. These scattering mechanisms can be sorted out into scattering by points S_p, lines S_L, surfaces S_S and intrinsic phonon–phonon scattering $S_{ph\text{-}ph}$. If the presence of one type of scattering mechanism does not affect another, then the total scattering rate S is simply a sum of the scattering rates (Matthiessen's Rule).

$$S = S_p + S_L + S_S + S_{ph\text{-}ph}. \tag{3}$$

2. Point scattering

Rayleigh scattering by isolated impurities, isotopes or point defects is generated by the difference in mass δM and modulus δG of the defect or impurity from the atoms of the host lattice (Ziman 1963).

$$S_p = \alpha q^4 (M/\rho)^2 [(\delta M/M)^2 + 3(\delta G/G)^2], \tag{4}$$

where α is a constant, q is the wave vector of the lattice wave or phonon ($q = 2\pi/\lambda$, where λ is the wavelength of the lattice wave) and ρ is the effective density of the impurity. Lattice distortions from either a contraction inwards in case of a small or an expansion outwards in the case of a large impurity surprisingly cause no scattering of lattice waves in an elastically isotropic solid (Ziman 1963). Diamond has an anisotropy factor of only 1.2 and approximates an elastically isotropic lattice (Kittel 1960).

Only boron, nitrogen and hydrogen dissolve in diamond (Field 1979). Hydrogen occupies a substitutional site in CVD diamond where it forms a sp^3 covalent bond with a neighbouring carbon atom (Baba 1991). Because diamond is a tetravalently bonded solid, the other three carbon atoms around this hydrogen have unsatisfied bonds and will form, respectively, three double bonds with neighbouring carbon atoms. These trigonally bonded carbons will try to force their bonds to be coplanar (Solomons 1980) and will change the local modulus around each hydrogen atom.

Substitutional hydrogen thus will scatter phonons and decrease the thermal conductivity as has been observed (Baba *et al.* 1991). The hydrogen content C_H of CVD diamond increases exponentially with increasing hydrocarbon concentration in the CVD gas mixture (Baba *et al.* 1991). The hydrocarbon concentration is a proxy for the diamond growth rate G because the diamond growth rate is a linear function of the hydrocarbon concentration in the gas. Hence,

$$C_H = C_{Ho} \exp[\gamma G], \tag{5}$$

where C_{Ho} is a constant and γ is a parameter dependent on atomic hydrogen concentration in the gas. Diamond grows by hydrocarbon radicals attaching to the diamond surface. Atomic hydrogen then abstracts away the hydrogen on the radical leaving behind pure diamond. If the diamond growth rate is high, not all of the hydrogen is removed and hydrogen becomes incorporated into the diamond. For processes such as the hot-filament CVD process, the atomic hydrogen concentration is low and γ is large. In high-temperature plasma processes, the concentration of atomic hydrogen is high and γ is small. Additions of oxygen can also lower γ.

Atomic hydrogen also stabilizes sp^3 bonds on the growth surface and prevents surface reconstruction. Low concentrations of atomic hydrogen and high growth rates may result in partial surface reconstruction and incorporation of the associated sp^2 bonds into the diamond bulk. sp^2 bonding decreases the thermal conductivity (Ono *et al.* 1986). The concentration C_{sp^2} of these sp^2 bonds will follow a relation similar to equation (5). The total scattering is

$$S_p \propto C_H + C_{sp^2} \propto \eta \exp[\gamma G], \tag{6}$$

where η is a constant. In diamond when sp^2 bonds and hydrogen are the dominant scatterers, equations (2) and (6) yield the thermal conductivity K as a function of the diamond growth rate G,

$$K = \beta/S_p = (\beta/\eta) \exp[-\gamma G]. \tag{7}$$

Equations (2) and (5) predict a hyperbolic relation between the thermal conductivity and the hydrogen content of the diamond. This is the type of relationship observed by Baba *et al.* (1991). Equation (7) indicates that films with the highest thermal conductivity are grown at the lowest diamond growth rate G in agreement with experiment (Graebner *et al.* 1992*a*). In addition, films with the highest thermal conductivity are grown with the most atomic hydrogen or with oxygen additions, i.e. low γ.

The second most prevalent 'impurity' in CVD diamond is the isotope ^{13}C at a concentration of 1.1% in a matrix of 98.9% ^{12}C. The ^{13}C atom is slightly smaller than the ^{12}C atom because the heavier isotope has a lower zero-point energy (Holloway *et al.* 1991). As discussed above, this size difference will not scatter phonons. Since the moduli are equal, only the mass difference causes phonon scattering. Pomeranchuk (1942) first predicted that isotopically pure dielectrics would conduct heat better than the same substances containing their natural abundance of isotopes. Recently, isotopically pure single-crystal diamond was found to have a thermal conductivity 50% higher than that of natural isotopic abundance diamond at room temperature (Anthony *et al.* 1990). This large enhancement has been theoretically explained in different ways (Berman *et al.* 1976; Nepsha *et al.* 1991; Bray *et al.* 1991; Hass *et al.* 1992; Berman 1992). The thermal conductivity of diamond has been measured across the entire ^{12}C–^{13}C binary diagram (Anthony *et al.* 1992) and at high and low temperatures (Onn *et al.* 1992).

The first attempt to see an isotope effect in a CVD diamond film failed because other mechanisms dominated isotope scattering (Anthony *et al.* 1991). High-quality CVD diamond films with thermal conductivities of 17 W cm^{-1} K^{-1} (Morelli *et al.* 1991) would have a thermal conductivity of 25 W cm^{-1} K^{-1} if they were isotopically pure (Anthony *et al.* 1992).

Other common impurities in CVD diamond films are metals that are contaminants from hot filaments or electrodes. In addition to causing phonon scattering, these metals can also induce the formation of sp^2 bonds in diamond similar to the case of hydrogen. The rate of incorporation I_M of metal in diamond from a hot filament increases exponentially with the filament temperature (Hinneberg 1992),

$$I_M = I_0 \exp(-H/kT), \tag{8}$$

where I_0 and H are constants that depend on the filament. Because the concentration of carbon atoms is much greater than metal atoms, the concentration per unit volume C_M of metal in a diamond film growing at a rate of G is

$$C_M = I_M/G. \tag{9}$$

If both hydrogen and metal dominate phonon scattering in the growing diamond film, the scattering rate S_p is

$$S_p \propto C_H + \phi C_M = C_{Ho} \exp[\gamma G] + \phi[I_M]/G, \tag{10}$$

where ϕ is the phonon scattering ratio of a heavy metal relative to a hydrogen atom. Equation (10) can be differentiated with respect to G to find a minimum in phonon scattering ratio as a function of the diamond growth rate. At high growth rates, hydrogen incorporation dominates and causes high rates of phonon scattering. At low diamond growth rates, metal incorporation prevails and also causes high rates of phonon scattering. At intermediate diamond growth rates, the scattering reaches a minimum and the thermal conductivity attains its maximum value.

Vacant sites also decrease the thermal conductivity because a vacancy has a large modulus change ($\delta G = G$) and a large mass change ($\delta M = M$). The equilibrium concentration of vacancies is small because of the high energy of formation of vacancies in diamond. Even so, in highly perfect isotopically pure diamonds, effects of vacancies on the thermal conductivity of diamonds may be seen (Onn *et al.* 1992). In CVD diamond processes, high concentrations of vacancies may be grown into the crystal because of lack of atomic mobility. High diamond growth rates may aggravate this situation.

3. Line scattering

Phonons can be scattered by dislocations by three mechanisms. First, the long-range asymmetric strain field around a dislocation scatters phonons. This effect dominates the second mechanism, namely, scattering by the core of the dislocation. The third scattering mechanism is a dynamic one in which mobile dislocations 'flutter' in the phonon breeze and is not important in diamond at room temperature. The phonon scattering S_{dis} for a dislocation density of N_{d} perpendicular to the phonon propagation direction is

$$S_{\text{dis}} \propto N_{\text{d}} \Gamma b^2 q, \tag{11}$$

where Γ is Gruneisen's constant, b is the Burgess vector of the dislocation and q is the wave vector of the phonon (Ziman 1963).

Most dislocations in diamonds lie parallel to the crystal growth direction (Wilks *et al.* 1991). Hence, dislocations in CVD films are oriented in a direction perpendicular to the substrate. Dislocations parallel to the direction of phonon motion scatter phonons weakly while dislocations perpendicular to the direction of phonon motion scatter phonons strongly (Klemens 1958; Ziman 1963). Consequently, the thermal conductivity of CVD films parallel to the surface is less than the thermal conductivity perpendicular to the surface. This type of behaviour has been observed in various diamond films (Graebner *et al.* 1992*a*).

Competitive growth between adjacent grains develops a cone-like columnar grain structure perpendicular to the substrate in CVD diamond films (Wild *et al.* 1989, 1990). Certain favourably oriented grains dominate and grow larger so that the average grain size increases monotonically from the substrate to the growth side of the film. Dislocations within each grain fan out as the grains grow in cross sectional area (Wilks *et al.* 1991) so that the dislocation density N_{d} decreases as the inverse square of the distance Z from the substrate side of the film,

$$N_{\text{d}} = N_0/Z^2, \tag{12}$$

where N_0 is the dislocation density near the substrate surface. A second scattering mechanism will eventually dominate when the dislocation density falls below a critical value. Addition of this second phonon scattering mechanism and combination of equations (2), (11) and (12) show that the thermal conductivity $K(Z)$ in the film increases from the substrate to the growth surface where it approaches a constant value as

$$K(Z) \propto (C_1 Z^2)/(Z^2 + C_2), \tag{13}$$

where C_1 and C_2 are constants. This qualitative type of behaviour has been observed by Graebner *et al.* (1992*b*) in some CVD films.

4. Surface scattering

If CVD films are extremely thin or at low temperatures, phonon scattering by external surfaces may affect the thermal conductivity of the films by effectively limiting the mean free path Λ to the film thickness Z_0. Substitution into equation (1) gives (Klemens 1958):

$$K = \tfrac{1}{3} C_v V Z_0. \tag{14}$$

Almost all CVD films are polycrystalline films with a cone-like columnar structure with the grain size increasing from the substrate to the growth surface. The phonon scattering power of simple grain boundaries in diamonds should not be large because of the destructive interference of long range strain fields by the multiple dislocations at boundaries. Scattering comes only from dislocation cores which are ineffective scatters of the type of long wavelength (1500 atom spacings) phonons that are dominant in diamond at room temperature. Consequently, phonons impinging on boundaries at normal incidence are only weakly scattered. Only phonons impinging on boundaries at glancing angles will be strongly scattered (Ziman 1963). However, such glancing collisions make up only a minority of the phonon–boundary collisions.

It is possible that the width of grain boundaries and thus the degree of boundary phonon scattering in CVD films exceeds the normal boundary parameters because of hydrogen segregation or dislocations clustering near grain boundaries. Then, the thermal conductivity for CVD diamond with a grain size of D would be (Klemens 1958):

$$K = \begin{cases} C_v V \Lambda_i, & \Lambda_i < D, \quad (15a) \\ C_v V D, & \Lambda_i > D, \quad (15b) \end{cases}$$

where Λ_i is the intrinsic mean free path of phonons in the diamond found within the CVD grain. For high quality diamond with a thermal conductivity of 22 W cm^{-1} K^{-1} at room temperature, $\Lambda_i = 0.2$ μm. Most CVD diamond films 300–500 μm thick have an average grain size $\langle D \rangle$ of at least 20 μm which is 100 times greater than the intrinsic mean free path of phonons Λ_i. This condition $\Lambda_i \ll D$ implies that grain boundaries in these CVD films should not cause anisotropies or gradients in thermal conductivities of the type observed by Graebner *et al.* (1992*a–c*).

For polycrystalline CVD films with a much smaller columnar grain size, gradients and anisotropy in the thermal conductivity are possible. In this case, a derivation similar to equation (13) shows that $K(Z)$ varies as,

$$K(Z) \propto (C_3 Z)/(Z + C_4), \quad \Lambda_i > D. \tag{16}$$

5. Temperature dependence

The temperature dependence $K(T)$ of the thermal conductivity of CVD diamond films (Morelli *et al.* 1988, 1991*b*; Anthony *et al.* 1991) is similar to previous work on diamond gem stones (Berman *et al.* 1976). Both the temperature dependence of the specific heat $C_V(T)$ and the dependence of the scattering factor S on the phonon wavevector $q(T)$ which varies with temperature affect $K(T)$. Up to room temperature, $C_V(T)$ follows a T^3 dependence (Kittel 1960).

The peak in the phonon energy content at a temperature T occurs at a phonon frequency equivalent to $4T$ (Klemens 1981). Scattering of these phonons has the most effect on the thermal conductivity. The wave vector q of these phonons is

$q(T) = 8\pi kT/(hV)$, where k is the Boltzmann and h is the Planck constant. For point, line and surface phonon scattering, $K(T)$ is from equations (2), (4), (11) and (15) (Klemens 1958):

$$K(T) \propto C_v(T)/S(T) \propto \begin{cases} T^3/T^4 = 1/T, & \text{point scattering,} \quad (17a) \\ T^3/T^1 = T^2, & \text{line scattering,} \quad (17b) \\ T^3/T^0 = T^3, & \text{surface scattering.} \quad (17c) \end{cases}$$

Morelli *et al.* (1988, 1991*b*) and Anthony *et al.* (1991) have seen a $T^{1.9-2.2}$ dependence of the thermal conductivity at low temperatures suggesting that dislocations are the dominant scatterer.

For high conductivity films, non-conservative phonon–phonon scattering (Umklapp processes) dominate at temperatures above the conductivity maximum at 150–170 K. Umklapp processes that involve very energetic phonons degrade the thermal conductivity because they do not conserve momentum. The energy of Umklapp phonons is approximately $\frac{1}{2}k\Theta$, where Θ is the Debye temperature (1840 K). The number of these phonons is given by the Boltzmann factor, $\exp(-\Theta/2T)$. The scattering factor $S_{\text{ph-ph}}$ is proportional to the square of the number of these energetic phonons. Combination of equations (2) and (3) yields $K(T)$ of diamond above conductivity maximum:

$$K(T) \propto T^3 \exp(\Theta/T), \quad \text{phonon–phonon scattering.} \qquad (18)$$

References

Anthony, T. R., Banholzer, W. F., Fleischer, J. F., Wei, L., Kuo, P. K., Thomas, R. L. & Pryor, R. W. 1990 Thermal diffusivity of isotopically enriched ^{12}C diamond. *Phys. Rev.* B **42**, 1104–1111.

Anthony, T. R., Fleischer, J. F., Olson, J. R. & Cahill, D. G. 1991 The thermal conductivity of isotopically enriched polycrystalline diamond films. *J. appl. Phys.* **69**, 8122–8125.

Anthony, T. R. & Banholzer, W. F. 1992 Properties of diamond with varying isotopic composition. *Diamond Related Mater.* **1**, 717–726.

Baba, K., Aikawa, Y. & Shohata, N. 1991 Thermal conductivity of diamond films. *J. appl. Phys.* **69**, 7313–7315.

Berman, R. & Martinez, M. 1976 The thermal conductivity of diamond. In *Diamond Research*, pp. 7–13. (Suppl. to *Ind. Diamond Rev.*)

Berman, R. 1992 The thermal conductivity of isotopically enriched diamonds. *Phys. Rev.* B **45**, 5726–5728.

Bray, J. W. & Anthony, T. R. 1991 On the thermal conductivity of diamond under changes to its isotopic character. *Z. Phys.* B*: Condensed Matter* **84**, 51–57.

Field, J. E. 1979 *The properties of diamond.* London: Academic Press.

Graebner, J. E., Mucha, J. A., Seibles, L. & Kammlott, G. W. 1992*a* The thermal conductivity of CVD diamond films on silicon. *J. appl. Phys.* **71**, 3143–3146.

Graebner, J. E., Jin, S., Kammlott, G. W., Herb, J. A. & Gardinier, C. F. 1992*b* Unusually high thermal conductivity in diamond films. *Appl. Phys. Lett.* **60**, 1576–1578.

Graebner, J. E., Jin, S., Kammlott, G. W., Bacon, B., Seibles, L. & Banholzer, W. F. 1992*c* Anisotropic thermal conductivity in CVD diamond. *J appl. Phys.* **71**, 5353–5356.

Hass, K. C., Tamor, M. A., Anthony, T. R. & Banholzer, W. F. 1992 Lattice dynamics and Raman spectra of isotopically mixed diamond. *Phys. Rev.* B **45**, 7171–7182.

Hinneberg, H. J., Eck, M. & Schmidt, K. 1992 Hot-filament-grown diamond films on Si: characterization of impurities. *Diamond Related Mater.* **1**, 810–813.

Holloway, H., Hass, K. C., Tamor, M. A., Anthony, T. R. & Banholzer, W. F. 1991 Isotopic Dependence of the Lattice Constant of Diamond. *Phys. Rev.* B **44**, 7123–7126.

Kittel, C. 1960 *Introduction to solid state physics.* New York: John Wiley & Sons.

Klemens, P. G. 1958 Lattice thermal conductivity. *Solid State Phys.* **7**, 1–98.

Klemens, P. G. 1981 Thermal conductivity of pure monoisotopic silicon. *Int. J. Thermophys.* **2**, 323–330.

Morelli, D. T., Beetz, C. P. & Perry, T. A. 1988 Thermal conductivity of synthetic diamond films. *J. appl. Phys.* **64**, 3063–3068.

Morelli, D. T., Smith, G. W., Heremans, J., Banholzer, W. F. & Anthony, T. R. 1991*a* Thermal properties of diamond single crystals with varying isotopic composition. In *New Diamond Science and Technology* (ed. R. Messier, J. T. Glass, J. E. Butler & R. Roy), pp. 869–873. Pittsburgh: Materials Research Society.

Morelli, D. T., Hartnett, T. M. & Robinson, C. J. 1991*b* Phonon-defect scattering in high thermal conductivity diamond films. *Appl. Phys. Lett.* **59**, 2112–2114.

Nepsha, V. I., Grinberg, V. R., Klyuyev, Y. A., Naletov, A. M. & Bokii, G R. 1991 The role of elastic interactions of phonons in diamond. *Dokl. Akad. Nauk. SSSR* **317**, 96–97.

Onn, D. G., Witek, A., Qiu, Y. Z., Anthony, T. R. & Banholzer, W. F. 1992 Some aspects of the thermal conductivity of isotopically enriched diamond single crystals. *Phys. Rev. Lett.* **68**, 2806–2809.

Ono, A., Baba, T., Funamoto, H. & Nishikawa, A. 1986 *Jap. J. appl. Phys.* **25**, L808–L810.

Pomeranchuk, I. 1942 On the thermal conductivity of dielectrics at temperatures lower than that of Debye. *J. Phys. USSR* **6**, 237–250.

Solomons, T. W. G. 1980 *Organic chemistry* 2nd edn. New York: John Wiley & Sons.

Wild, Ch., Herres, N., Wagner, J., Koidl, P. & Anthony, T. R. 1989 Optical and structural characterization of CVD diamond. *Proc. electrochem. Soc.* **89** (12), 283.

Wild, Ch., Herres, N. & Koidl, P. 1990 Texture formation in polycrystalline diamond films. *J. appl. Phys.* **68**, 973–978.

Wilks, J. & Wilks, E. 1991 *Properties and applications of diamond.* Oxford: Butterworth–Heinemann.

Ziman, J. M. 1963 *Electrons and phonons: the theory of transport phenomena in solids.* Oxford: Clarendon Press.

7

Electron irradiation and heat treatment of polycrystalline CVD diamond

BY C. D. CLARK AND C. B. DICKERSON

Photoluminescence and Raman spectra have been used to characterize the properties of diamond films grown by microwave plasma assisted chemical vapour deposition. Measurements at 77 K and excitation wavelengths in the range 476.5 nm to 514.5 nm show the presence of two components, A and B, in the Raman spectrum in addition to the diamond Raman line. The A and B components are rather similar in appearance and show resonant Raman behaviour. Electron irradiation results in the removal of the A and B Raman components, but they return to their original strength after heating at 600 °C. The Raman scattering species interact with other point defects in the CVD films during heat treatment, and may be related to the presence of silicon in the diamond film.

1. Introduction

Raman spectroscopy is often applied to the characterization of diamond films produced by chemical vapour deposition (CVD) because diamond has a strong sharp first order Raman line at 1332 cm^{-1} which is distinguished easily. A variety of other 'non-diamond' Raman features have been reported for CVD diamond, in addition to the 1332 cm^{-1} diamond Raman line. In principle the energies of the vibrational modes involved together with their associated Raman intensities can be used to assess the quality and purity of the diamond films being produced by different growth procedures. For this to be effective a good understanding of the nature of the Raman scattering species and their concentrations in the film are required.

The technique of photoluminescence in diamond is useful as a means of studying the behaviour of point defects in CVD diamond films because there is a substantial coverage of defect properties already in the scientific literature (see, for instance, the review article by Davies (1977) and the book edited by Field (1992)). In the present investigation measurements of Raman scattering and photoluminescence have been made simultaneously and an in-depth characterization study of three CVD diamond films has been carried out.

Figure 1 shows the emission spectra of three polycrystalline diamond films grown by microwave assisted CVD onto a silicon substrate. The spectra were recorded at 77 K under illumination with 514.5 nm light. The normalized emission spectra were obtained by dividing the recorded spectrum by the intensity of the first-order diamond Raman line at 1332 cm^{-1}. This procedure facilitates semiquantitative comparison of spectra recorded on the same fragment of sample after any electron irradiation or thermal annealing treatment. All the spectra presented in this paper

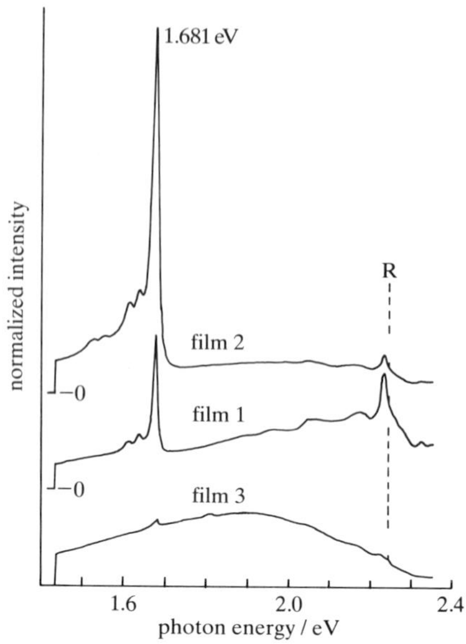

Figure 1. Emission spectra recorded at 77 K and excited with 514.5 nm light for CVD polycrystalline diamond films. The first order Raman line for diamond is indicated by R. The spectra have been displaced vertically for clarity of presentation. The zero levels of intensity are indicated.

have been corrected for the wavelength dependent response of the optical measuring system.

There are three major features of the spectra other than the first-order diamond Raman line at 1332 cm^{-1}, and they are present in different proportions in the three samples:

(i) a zero-phonon line at 1.681 eV with its weak vibronic sideband structure;

(ii) a Raman scattering spectrum between 2.0 and 2.4 eV, being strongest in film 1 and showing detailed structure which is much better resolved at 77 K than at room temperature;

(iii) a broad luminescence band at *ca.* 1.90 eV, being of greatest intensity in film 3.

The effects of heat treatment up to 2200 °C and of electron irradiation followed by heat treatment on the 1.681 eV centre for film 2 have been fully described in an earlier publication (Clark & Dickerson 1991). It has been established that the 1.681 eV centre is associated with silicon as an impurity in the diamond film.

A strong resonant Raman scattering behaviour also has been reported previously (Clark & Dickerson 1992) for the 'non-diamond' Raman spectrum. Two Raman components were identified and labelled A and B. The intensity of the B component was very sensitive to the wavelength of the exciting light in the range 476.5–514.5 nm. Only the A Raman component was present in the spectrum of film 3 and its normalized intensity increased as the excitation changed from 476.5–514.5 nm. In film 2 the B Raman component showed a substantial increase in intensity as the excitation wavelength increased, reaching about twice the intensity of the A

Figure 2

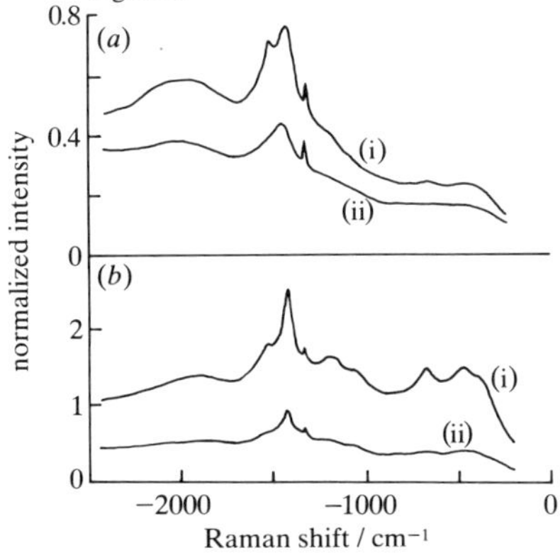

Figure 3

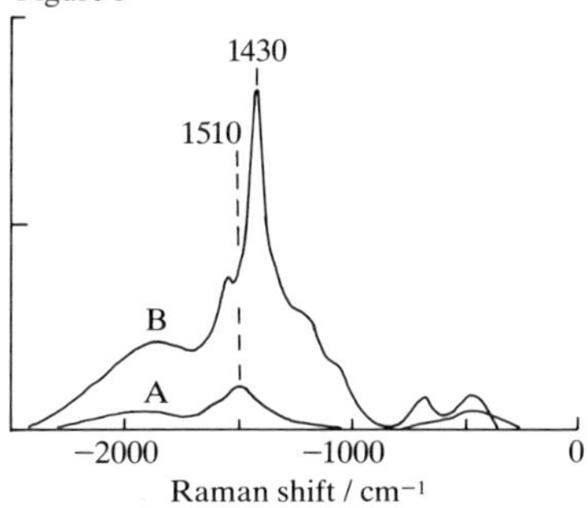

Figure 2. Raman shift spectra of film 2, recorded at 77 K and excited with (*a*) 488.0 or (*b*) 514.5 nm light, after heat treatments at (i) 1000 °C and (ii) 1300 °C.

Figure 3. Deconvoluted A and B Raman spectra for CVD film 2 at 77 K.

component under 514.5 nm excitation. Both Raman components seem to bear a strong resemblance to the Raman spectra of amorphous $Si_x C_{1-x}$ reported by Ramsteiner *et al.* (1988), and because of this have been associated at least in part with the presence of silicon in the diamond film.

Clark & Dickerson (1992) also have reported that the broad band luminescence at 1.90 eV and its variation in intensity in the temperature range 72–300 K has a strong resemblance to the photoluminescence behaviour of amorphous hydrogenated $Si_x C_{1-x}$ alloys as reported by Sussmann & Ogden (1981).

The purpose of this paper is to report the further behaviour of the same three diamond films reported in our earlier publications, as a result of electron irradiation and heat treatment.

2. Results

(*a*) *Heat treatment of as-grown CVD diamond film*

A small fragment of the as-grown film 2 has been subjected to a series of isochronal heat treatments for thirty minutes under high vacuum conditions in the temperature range 800–1350 °C. The Raman spectra excited by 488.0 and 514.5 nm light after the 1000 and 1300 °C anneals are shown in figure 2. The spectra have been normalized, but note that the 514.5 nm spectra are more intense than the 488 nm spectra. From 800–1200 °C some variations of normalized intensity were observed for both excitations but the general shapes of the two sets of spectra did not show much change. Up to 1200 °C each of the spectra could be deconvoluted quite well into the A and B Raman components shown in figure 3. These A and B components are refinements to the original A and B spectra resulting from the fitting of a wider range of spectra. Figure 2 shows that under 488.0 and 514.5 nm excitation there is a sharp increase in the measured Raman intensity after the 1300 °C heat treatment. With the exception of an obvious additional peak at *ca.* 1600 cm^{-1} the shape of the 488 nm spectrum was not appreciably altered. For the 514.5 nm excitation the relative intensities of the spectral features changed. Thus the features at 450, 670, 1070 and 1185 cm^{-1} in the 514.5 nm Raman shift spectrum showed a greater increase in

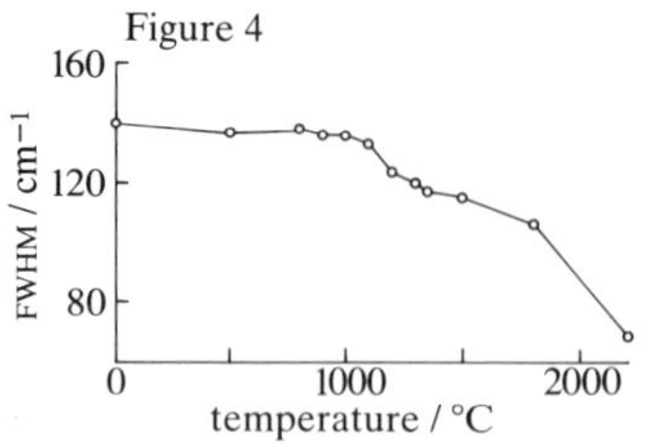

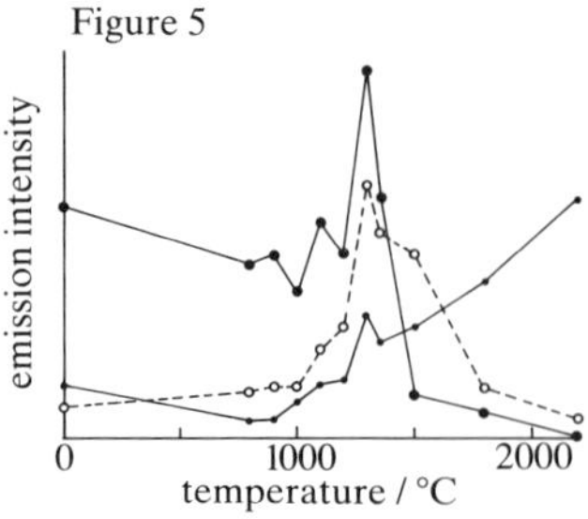

Figure 4. The change in the half-width (FWHM) of the 1.681 eV zero-phonon line in film 2 as a function of the annealing temperature. Measurements of half-width were recorded at 77 K.

Figure 5. The changes in emission intensity of the Raman signal at 1430 cm^{-1} (large spots), the 1.681 eV zero-phonon line (small spots) and the 1.967 eV zero-phonon line (open circles) for film 2 as a function of thermal annealing temperature. Spectra were recorded at 77 K using 514.5 nm exciting light.

strength than the peaks at higher energies. The 514.5 nm spectrum no longer could be described solely in terms of the A and B Raman components and an additional Raman spectrum appeared to have been produced.

The sharpening up of the peaks at 450, 670, 1070 and 1185 cm^{-1} suggests that the film may have been relieved of some of the strain which had been introduced during growth. Because of the large half widths and possible overlaps between these features it is difficult to obtain a quantitative measure of these linewidth changes. However, the 1.681 eV line is much sharper and shows a 13% reduction in linewidth between 1000 and 1350 °C as shown in figure 4.

The isochronal annealing has been extended to the temperature range 1500–2200 °C, but for these treatments the samples were subjected also to a hydrostatic pressure of 9 GPa to maintain diamond stability. At 1350 °C the Raman intensity of both A and B components showed a sharp reduction in strength and this was followed by a further reduction at 1500 °C. The overall annealing behaviour of the Raman scattering is shown in figure 5, where the background corrected Raman scattering signal at 1430 cm^{-1} (recorded under 514.5 nm excitation) is plotted as a function of the annealing temperature. The signal at 1430 cm^{-1} indicates the behaviour of the strongest feature in the Raman spectra, but all the other features exhibit a very similar behaviour. There are fluctuations in intensity up to 1200 °C which may be due to experimental uncertainties in the normalization procedures, but there is a clear peak in intensity at 1300 °C followed by a sharp reduction at 1500 °C.

During the isochronal heat treatment of as-grown films 1 and 2 a zero-phonon line was observed at 1.967 eV. A much weaker line at 1.923 eV is probably an associated, phonon-assisted transition, indicating a rather weak coupling to the lattice, rather similar to the 1.681 eV centre. The variations in the integrated intensities of the 1.967 eV and 1.681 eV zero-phonon lines with annealing temperature are compared with the Raman scattering behaviour in figure 5. All three features show an abrupt change in behaviour at 1300 °C indicating that the events are linked in some way. It is suggested that all three features may be influenced by the presence of silicon impurity in the films. As the heating sequence proceeds it appears that the silicon present in the film is converted into the form of the 1.681 eV centre.

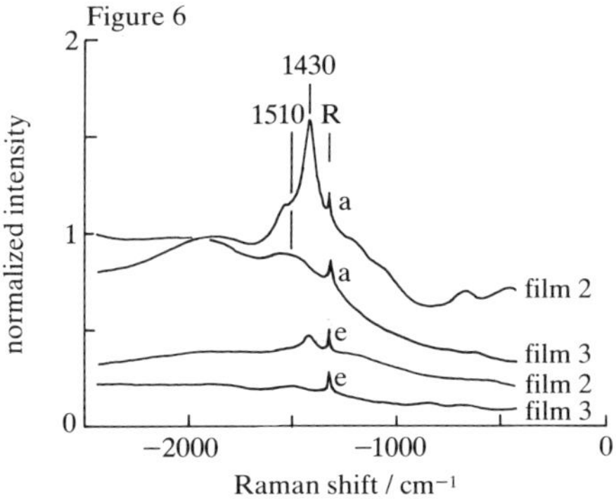

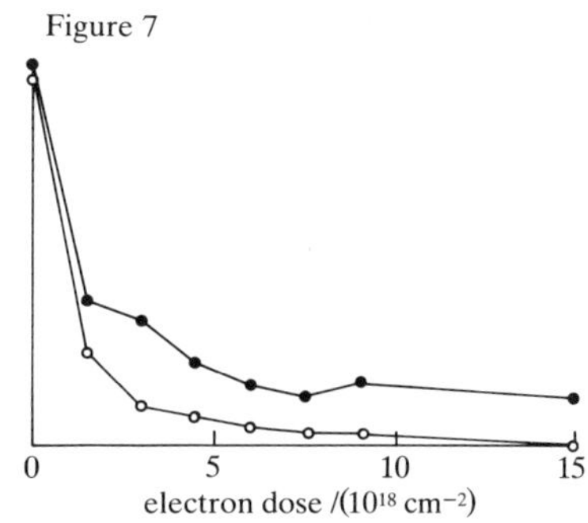

Figure 6. Raman shift spectra recorded at 77 K for films 2 and 3 in the as-grown condition (labelled a) and after electron irradiation (labelled e).

Figure 7. The Raman signal at 1500 cm^{-1} for CVD diamond film 2 as a function of 2.0 MeV electron dose (open circles). The emission was recorded at 77 K using 514.5 nm exciting light. The dots show the electron dose dependence of the signal at 1.90 eV for film 3 using 514.5 nm exciting light.

(*b*) *Electron irradiation of as-grown* CVD *diamond*

Small fragments of films 2 and 3 were irradiated at room temperature with 2.0 MeV electrons and the emission spectra excited with 514.5 nm light were measured as a function of the electron dose. The spectra in figure 6 were recorded before and after irradiation. That both the A and B Raman components were reduced in intensity can be seen because film 3 contains almost exclusively component A, whereas film 2 contains both the A and B components in roughly the same relative proportions.

Figure 7 shows the electron dose dependence of:

(*a*) the background corrected Raman signal at 1500 cm^{-1} for 514.5 nm excitation of film 2 (this parameter includes contributions from both the A and B Raman components);

(*b*) the photoluminescence signal at 1.90 eV for film 3.

It is apparent that both the A and B Raman scattering features and the broad band photoluminescence at 1.90 eV are substantially removed by an irradiation dose of 2×10^{18} electrons cm^{-2}.

(*c*) *Heat treatment of electron irradiated* CVD *film*

The changes in the Raman shift spectrum of an electron irradiated fragment of film 1 were recorded after a series of heat treatments and the results are summarized in figure 8. The spectra shown were recorded under 488 nm excitation. After the 300 °C anneal there appears to be a sharpening of the individual Raman components and, at the same time, there is an overall decrease in signal intensity. For heat treatments above 300 °C the Raman scattering signal increases in intensity and after heating at 600 °C the spectrum had recovered to be very similar to the intensity and form of the spectrum of the as-grown sample. A very similar set of results was obtained for film 2 under 514.5 nm excitation, so that it is concluded that both the A and B Raman components respond in the same way to electron irradiation and subsequent heat treatment.

It is not easy to investigate the thermal annealing behaviour of the 1.90 eV broad

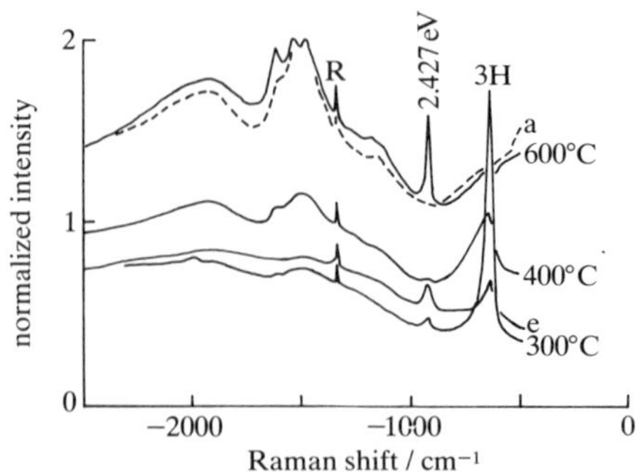

Figure 8. Raman shift spectra of film 1 in the as-grown condition (broken line), after electron irradiation (labelled e) and after thermal annealing at 300 °C, 400 °C and 600 °C. All the spectra were recorded at 77 K using 488.0 nm exciting light.

band luminescence because in film 3 the 2.156 eV centre (a centre thought to involve nitrogen impurity in association with a vacant lattice site) with strong vibronic sideband is generated as the vacancies become mobile, and this emission strongly overlaps with the 1.90 eV band. However, after heating at 1200 °C the 2.156 eV luminescence was no longer present in the emission spectrum and at this stage a broad band centred at 1.90 eV was observed at an intensity 1.4 times the intensity of the band observed in the as-grown condition. This behaviour is very similar to that of the Raman scattering.

3. Discussion

Irradiation of natural diamond with 2.0 MeV electrons usually produces a zero-phonon line at 1.673 eV with a strong vibronic sideband and it is generally accepted to be due to the neutral vacancy. It is often referred to in the literature as the GR1 centre. This line has been induced in the CVD diamond films studied in the present investigation but it is in close proximity to the 1.681 eV zero-phonon line when it too is present in the spectrum. These lines can be distinguished provided they are not seriously strain broadened and the absorption or luminescence measurements are carried out at low temperatures. The two lines also can be distinguished because they have very different vibronic sideband structures.

Since vacancies act as donor centres when introduced into the p-type semi-conducting diamonds measurements of the changing carrier concentration in semiconducting diamond can provide an estimate of the rate of production of intrinsic defects in diamond during electron irradiation. Clark *et al.* (1961) found that during 2.0 MeV electron irradiation the number of donor levels produced per incident electron was about 10. Since the depth of penetration of 2.0 MeV electrons in diamond is close to 2 mm, a dose of 2×10^{18} electrons cm^{-2} would be expected to produce a concentration of 10^{20} defects cm^{-3} in the surface layer. This corresponds to about 6 in 10^4 atoms being displaced. Similar proportions of displaced atoms would be expected for other non-ionically bonded bulk materials.

In view of the above discussion, since an electron dose of 2×10^{18} electrons cm^{-2} is sufficient almost to eliminate the A and B Raman components and the 1.90 eV band, it seems unlikely that they can be associated with bulk material. The complete recovery of the A and B Raman scattering by thermal annealing at 300 and 600 °C also is difficult to understand in terms of bulk material, but it could be explained more readily if the A and B Raman spectra are associated with defect-like species.

During the thermal annealing of the as-grown films, the A and B Raman components, the 1.681 eV centre and the 1.967 eV centre all show an abrupt annealing peak at 1300 °C. This seems to indicate that they are closely related to one another, and since the 1.681 eV centre is known to be a point defect in diamond, the other features are presumed also to be defects in diamond. In this respect it should be noted that silicon concentrations of between 15 and 50 ppm have been recorded in similar CVD diamond films by secondary ion mass spectroscopy (SIMS), and this is much lower than the estimated intrinsic defect concentration introduced by electron irradiation and described earlier.

The conclusion that the Raman scattering in the present work is associated with defects in CVD diamond implies that high Raman scattering cross sections are involved. It is not easy to propose models for the species which are responsible for the Raman scattering, but it is worth discussing their behaviour in relation to other more familiar point defects in diamond.

Vacancies in diamond are known to diffuse at temperatures in excess of 600 °C (Collins 1979) and they contribute to the formation of the 1.681 eV centre (Clark & Dickerson, 1991) in irradiated CVD diamond film. Since they are immobile below 600 °C vacancies would not seem to be involved in the Raman scattering behaviour below 600 °C.

There is no clear picture about the role of interstitial carbon atoms in diamond. Lomer & Marriott (1979) have suggested that interstitials might diffuse at temperatures as low as 50 K. It has further been speculated that mobile interstitials might become trapped at various sites from which they could be thermally released at different temperatures in the range up to 600 °C. Thus an involvement of interstitials with the Raman scattering centres might be a possibility.

Another mechanism which might affect the behaviour of the Raman scattering is charge redistribution amongst the defects present. In principle this can occur during electron irradiation or during optical illumination of the sample. In relation to this the defect known as 3H (with zero-phonon line at 2.462 eV, and illustrated in figure 8) has been reported to show charge exchange effects after irradiation and subsequent thermal annealing up to 600 °C (Palmer 1961). Figure 8 also shows the presence of a previously unreported defect with zero-phonon line at 2.427 eV which also exhibits reactivity during heat treatment up to 600 °C. A more detailed investigation of the removal and recovery of the Raman scattering spectrum is required.

Resonant Raman scattering behaviour in CVD diamond has been reported previously by Wagner *et al.* (1991) using a wide range of excitation photon energies (1.16–4.82 eV instead of 2.54–2.41 eV in the present study). The positions of peaks in their 290 K emission spectra depended upon the excitation photon energy, and the general shapes of the spectra were not the same as those reported here. It will be interesting to see how such features respond to the temperature at which they are measured, and to investigate their response to electron irradiation and heat treatment.

References

Clark, C. D. & Dickerson, C. D. 1991 The 1.681 eV centre in polycrystalline diamond. *Surf. Coatings Technol.* **47**, 336–343.

Clark, C. D. & Dickerson, C. D. 1992 Raman and photoluminescence spectra of as-grown diamond CVD films. *J. Phys. Condensed Matter* **4**, 869–878.

Clark, C. D., Kemmey, P. J. & Mitchell, E. W. J. 1961 Optical and electrical effects of radiation damage in diamond. *Discuss. Faraday Soc.* **31**, 96–106.

Collins, A. T. 1979 High temperature annealing of electron irradiated type I diamond. *Inst. Phys. Conf. Ser.* **46**, 327–333.

Davies, G. 1977 The optical properties of diamond. *Phys. Chem. Carbon* **13**, 1–143.

Field, J. 1992 *Properties of natural and synthetic diamond*. Academic Press. (In the press.)

Lomer, J. N. & Marriott, D. 1979 ESR in type IIa diamonds irradiated with electrons below 12 K. *Inst. Phys. Conf. Ser.* **46**, 341–346.

Palmer, D. W. 1961 Annealing of electron irradiation damage in diamond. Ph.D. thesis, University of Reading, U.K.

Ramsteiner, M., Wagner, J., Wild, C. & Koidl, P. 1988 Raman scattering of amorphous carbon/semiconductor interface layers. *Solid State Commun.* **67**, 15–18.

Sussmann, R. S. & Ogden, R. 1981 Photoluminescence and optical properties of plasma deposited amorphous Si_xC_{1-x} alloys. *Phil. Mag.* B **44**, 137–158.

Wagner, J., Wild, C. & Koidl, P. 1991 Resonance effects in Raman scattering from polycrystalline diamond films. *Appl. Phys. Lett.* **59**, 779–781.

Discussion

J. E. Butler (*Naval Research Laboratory, Washington, D.C., U.S.A.*): Raman features are observed with most of the characteristics of your A and B spectra in samples that do not show any evidence of the 1.681 eV photoluminescence feature attributed to a Si related defect. These were samples grown with an atmospheric oxygen–acetylene torch on molybdenum substrates. Your extensive analysis of these three Si containing CVD diamond films indicates interesting correlations of the non-1332 cm^{-1} Raman features (particularly in the 1340 to 1600 cm^{-1} region) with other point defect characteristics. Since the Raman spectra observed for various types and qualities of CVD diamond display similar features, please comment on the uniqueness of your deconvolution and the extention of your analysis to non-Si containing films.

C. D. Clark. In depth Raman scattering studies of the effects of heat treatment and electron irradiation need to be carried out on diamond films grown by other processes.

8

Strength, fracture and erosion properties of CVD diamond

BY J. E. FIELD, E. NICHOLSON, C. R. SEWARD AND Z. FENG

Theoretical and experimental studies have been made on the effect of high modulus coatings on the stress fields generated by indentation and impact onto a flat half-space. The theoretical work used finite-element techniques and it shows that a high modulus coating can have a significant effect on the maximum tensile stresses generated in the substrate providing there is a good bond at the coating/substrate interface. Because it is technically difficult to deposit layers of more than a few micrometres thickness without residual stresses causing debonding, double and multilayer systems have also been examined theoretically. A variety of techniques have been used to determine the strength, modulus, expansion coefficient, thermal conductivity and other physical properties of chemical vapour deposition CVD diamond layers. These are briefly reviewed and data from our own studies using such techniques as the vibrating reed, bulge test and indentation are given. The erosion properties of both CVD coated substrates and CVD free-standing layers are presented for both liquid drop and solid particle erosion. Finally, a study has also been made of the frictional properties of various CVD diamond layers in a range of environments; data are compared with our earlier work on bulk diamond.

1. Introduction

The strength properties of CVD (chemical vapour deposition) diamond films are of great current interest. Theoretical predictions, of the type outlined in §2, suggest that high modulus coatings should improve the strength, wear and erosion properties of components. However, questions arise about the strength of the interface and the size and nature of residual stresses left by the deposition process. To help with theoretical predictions, any assessment of the practical use of these coatings, and optimization of the CVD processes, it is important to measure properties such as hardness, moduli, Poisson ratio, thermal expansion coefficient and thermal conductivity; all, ideally, as functions of temperature. This paper briefly outlines techniques used for such measurements as well as methods for measuring the performance of CVD coatings in situations involving frictional rubbing, impact, erosion and wear.

2. Effect of a high modulus coating on contact stresses

A finite-element model has been developed by van der Zwaag & Field (1982) for modelling the effect of a high modulus coating on the stress field generated by hertzian contact (sphere on a flat). The coating layer was thin, being no more than 20% of the contact radius of the indent. It is shown that, as expected, the axial

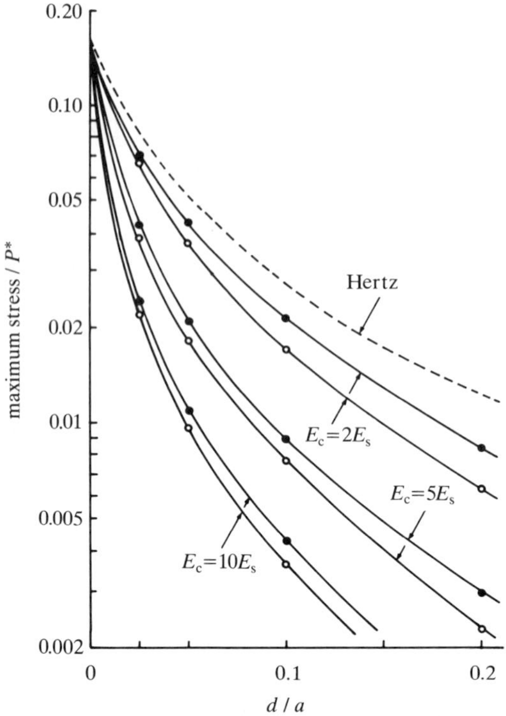

Figure 1. Variation of the maximum radial stress in the substrate with normalized coating thickness for a single layer coating; d is the coating thickness and a the radius of contact. The broken line is the Hertz solution for an isotropic solid (after van der Zwaag & Field 1982). ●, $\nu_c = 0.2$; ○, $\nu_c = 0.3$.

stresses are increased by the presence of the coating but the radial stresses (which can lead to tensile failure) are decreased. In this study, Young's modulus, Poisson's ratio and the thickness of the coating were varied. The results for the maximum tensile (radial) stress generated at the interface are summarized in figure 1. If the half-space were a uniform material then the stress would decrease with depth as shown by the dotted line. The decrease of stress at the depth of the interface, d, is increased dramatically by the high modulus coating; the examples given are for coating moduli, E_c, 2, 5 and 10 times the substrate modulus, E_s. (E_c/E_s for CVD diamond on silicon or zinc sulphide would be *ca.* 6 and 10 times respectively.) The major effects are with modulus and coating thickness, although the effect of Poisson's ratio is also significant. The physical reason that the high modulus coating reduces the build-up of the tensile radial stress is that it restricts the movement (displacement) of the substrate. For this to take place, a good bond between coating and substrate is essential.

As it is technically difficult to deposit layers with a thickness greater than a few micrometres without residual stresses resulting in debonding of the layer, double-layer systems where each layer has different elastic properties have also been investigated (van der Zwaag *et al.* 1986). The individual coating layers were taken as having equal thickness in this work. A combination of layers was found to be less effective than a thick layer with a modulus equivalent to the greater of the two moduli of the two coating layers. It was found that the order of these coating layers (i.e. whether the top or the bottom layer had the higher modulus) made no difference to the resultant stress in the substrate at the interface between the coating and the substrate. Again the effectiveness of the coating is reduced as the total coating thickness is reduced. There were differences, however, in both the stresses at the top surface of the coating and the interface between the coatings. The stresses in the interlayer are reduced, for example, when the higher modulus layer provides the outer layer of the coating. The surface stresses were again found to be increased for the coated material, the precise value depending on the combination chosen. Experimental work for germanium samples protected by hard carbon layers is given by van der Zwaag & Field (1983*a*).

In summary, it is clear that high modulus coatings have great potential, and also that finite element modelling can play a useful part. What restricts the modelling work at present is a lack of information on the physical properties of the coating, the strength of the bond between the coating and the substrate and the magnitude and sign of the residual stresses left by the deposition process. Important film properties which need to be measured are the hardness, moduli, the Poisson's ratio, the thermal expansion coefficient and the thermal conductivity; all, ideally, as functions of temperature.

3. Thin film property measurement

(*a*) *Hardness*

Typical Vickers micro-indenters are not suitable; however, nano-indenters, or ultramicro-indenters with very fine-scale indenter tips, are. A considerable amount of effort has been expended in interpreting the data and allowing the coating hardness to be evaluated (see, for example, Pethica *et al.* 1983; Isukamoto *et al.* 1987; Oliver & McHargue 1988; McHargue 1990; Smith *et al.* 1992).

(*b*) *Young's modulus*

Essentially four methods are used for this. First, a vibration technique in which a thin beam of substrate material of known properties is vibrated with and without a coating. The two resonant frequencies plus beam theory can be used to deduce the modulus of the coating (see, for example, Kinbara *et al.* 1981). Secondly, the 'blister' or 'bulge' technique (see, for example, Yang *et al.* 1977; Bromley *et al.* 1983; Allen *et al.* 1987; Cardinale & Tustison 1990, 1991) in which a circular area of substrate is etched away and the area of coating (now only supported about its periphery) is pressurized; the modulus can be calculated from the film deflection. Thirdly, by indentation with a nano-indenter and analysis of the displacement records (for references, see §3*a*). Finally, by Brillouin scattering using similar techniques and analysis to that used by Grimsditch & Ramdas (1975) for single crystal diamonds.

The first two methods are the cheaper and most direct solutions and can obtain values to a few percent accuracy provided the experimenter is careful enough and the films are of greater thickness than, say, 10 μm on a millimetre thickness substrate. A nano-indenter is a relatively expensive piece of equipment and the interpretation of the data for moduli is more complex. The Brillouin technique is potentially the most accurate but needs sophisticated equipment and skilled interpretation.

The modulus as a function of temperature can be measured using the vibrating

Figure 2

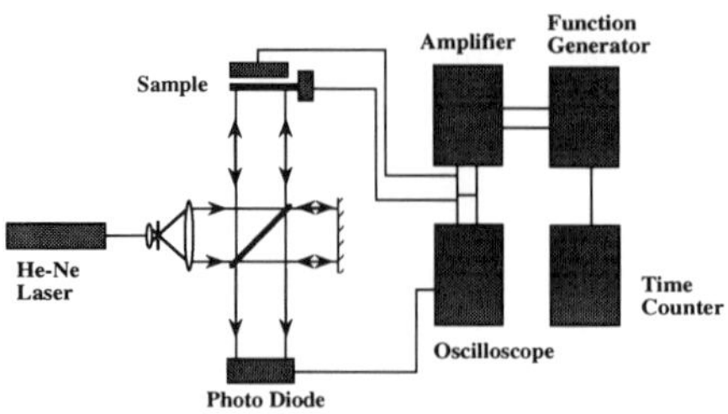

Figure 3

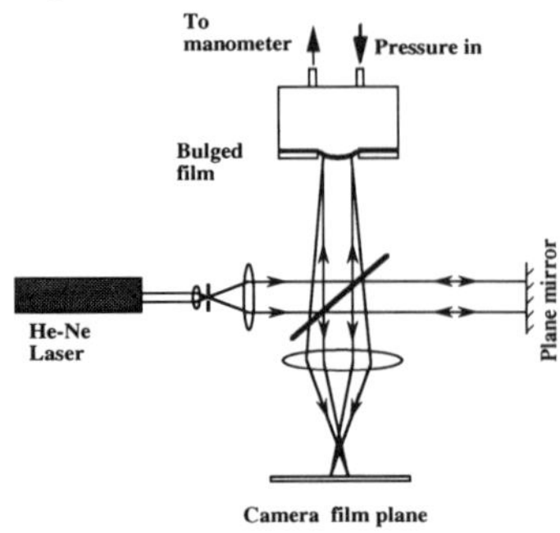

Figure 2. Schematic of the vibrating reed apparatus. At resonance the fringe visibility is lost and this is detected by the photodiode.

Figure 3. Schematic of the bulge test. Interference fringes allow the deflection of the membrane to be recorded.

need or 'blister' techniques provided a suitable furnace arrangement (with 'windows') is designed.

The vibrating reed apparatus used in our experiments is shown schematically in figure 2. The apparatus consists of two parts; vibration generation and frequency detection. The beams, of length 15–50 mm, width 3–5 mm are clamped to form part of a capacitor, with an aluminium reference electrode. A frequency function generator produces a variable frequency, sinusoidal voltage, which is amplified to a level required to cause the beam to vibrate. The vibrational frequency is detected using a Michelson's interferometer. A photodiode is placed behind a pin hole in the screen on to which the interference pattern falls. The pinhole is positioned at the edge of a bright fringe. At resonance, blurring out of the interference pattern occurs. This frequency is recorded and used to calculate the modulus of the beam.

In the case of very thin films, a composite beam is used. The modulus is calculated by first determining the resonant frequency of the substrate material (monolithic beam) and then repeating the experiment for the composite beam. It is from the shift in resonant frequency, that the Young's modulus of the film is calculated. In this way the modulus of deposited films as thin as a few tens of nanometres can be determined.

The Bulge test apparatus used in our experiments (figure 3) consists of a cylindrical chamber the front section of which has an opening of diameter greater than that of the membrane, through which the bulging of the film can be viewed. The sample is in the form of a 16 mm × 16 mm square of silicon coated with a hard thin film, with a circular membrane in the centre. A known pressure differential is applied across the membrane by means of a syringe. Pressures of up to 10^4 Pa (100 mbar) above atmosphere are produced within the chamber, which are measured using a manometer. A Michelson's interferometer is used to monitor the radius of curvature of the membrane. The central deflection of the membrane is found by counting the number of circular interference fringes formed as a function of pressure. Using this measurement, the radius of curvature of the membrane can be determined, from which the height of the bulge and hence the Young's modulus can be calculated. Because of diamond's extreme inertness to powerful etchants, an anisotropic etching technique, involving concentrated etchants at room temperature was used. The etchant used was a 1:1 mixture of concentrated HF (40%) and HNO_3 (Cardinale & Tustison 1990, 1991). The etch time for each sample varied from 5–10 min depending on the thickness of the silicon substrate. The diameter of the membranes were circular to within approximately 0.06 of a mm for a 6–6.5 mm diameter membrane.

(c) *Thermal conductivity (diffusivity)*

The thermal 'mirage' technique used by Anthony *et al.* (1990) to measure the thermal conductivities of isotopically pure diamond have been applied to thin film conductivity measurements (Pryor *et al.* 1990). The technique essentially involves heating a surface volume with one laser pulse and then interrogating the heated gas layers above the surface with a second laser beam travelling parallel to the specimen surface. The published data shows an almost linear increase of diffusivity with decreasing graphite content. Ono *et al.* (1986) have shown that the thermal conductivity of diamond films decreases as the methane to hydrogen ratio increases with decreasing graphite.

(d) *Coefficient of expansion*

This can in theory be measured for unsupported films, but if they are very thin (a few to tens of micrometres) then buckling is a major problem. It appears to be better to observe the deflections of coated substrates (with the substrate relatively thin). A similar furnace set up can be used as for the modulus versus temperature measurements, with beam deflections giving expansion coefficients and beam vibration resonances giving moduli.

(e) *Erosion properties*

Techniques are available at Cambridge for measuring the erosion resistance of materials to both solid particle and liquid drop impact. The solid particle impact apparatus uses gas flow to accelerate particles along a steel tube. The particles are fed into the flow from a reservoir using compressed air and the particle flux is controlled by means of a turntable in the reservoir. Impact velocities, which can be varied in the range up to *ca.* 300 m s^{-1}, are measured using a cross-correlation method to an accuracy of *ca.* 4%. Specimens can be mounted at any angle but normal incidence (i.e. 90°) was used in this work since this gives maximum erosion for 'brittle' materials. In the experiments on CVD films, the solid particles were sieved sand with dimensions of 300–600 μm and mean particle mass of 190 μg. The flux rate was 4.6 kg m^{-2} s^{-1} and impact velocities of 34 m s^{-1} and 59 m s^{-1} were chosen. Further details of the apparatus are given in Andrews *et al.* (1983) and Walley & Field (1987). Erosion data on bulk diamond and polycrystalline diamond composites (PDCs) are given in Hayward & Field (1990) and Feng & Field (1990).

Forward-facing aircraft components, particularly window materials, may suffer damage due to impact with rain drops. Materials coated with CVD diamond, or self-supporting CVD diamond layers themselves are currently under study as infrared transmitting windows.

Liquid impact studies have been in progress at Cambridge since the early work of Bowden & Brunton (1961) and Bowden & Field (1964). The key to understanding the pressure generated by liquid impact is the realization that there is always an initial regime where the contact edge velocity between liquid and target is higher than the shock wave velocities in the liquid and solid. This means that the liquid behind the shock envelope behaves compressibly, giving pressures of order ρCV, where ρ is the liquid density, V the impact velocity and C the shock velocity in the liquid. It is only when the shock envelope overtakes the contact edge that release waves move into the liquid allowing incompressible flow and sideways jetting (Bowden & Field 1964; Lesser 1981; Lesser & Field 1983).

For the impact velocities likely to arise in most rain erosion situations, the ratio of compressible to incompressible pressures is very large (34 times at 100 m s^{-1}, 10 times at 500 m s^{-1}). In other words, it is the very early stages of liquid impact which are all important. Impacts by rain drops of a few 100 m s^{-1} produce pressures of gigapascal magnitude and submicrosecond duration followed by lateral outflow (jetting) at a few times the impact velocity. The fact that the early stage of liquid impact is all-important also allows drop impact to be simulated by firing liquid jets at stationary targets (a much simpler approach than projecting specimens at suspended drops), though it is essential to produce jets that are coherent and that have a smooth, slightly curved front profile. It is possible to calibrate jet impacts in terms of equivalent drop sizes which they reproduce (Field *et al.* 1979).

The rain erosion testing described in this paper has been performed with a multiple impact jet apparatus (MIJA). This produces a series of reproducible jets of chosen dimension and in the velocity range up to *ca.* 600 m s^{-1}. A computer-controlled specimen stage allows impacts to be coincident or in arrays or random patterns that are reproducible. The computer also records all relevant impact parameters including each impact velocity (Seward *et al.* 1990).

In *quantitative* studies of liquid impact it is important to be able to determine 'threshold velocities' for damage, i.e. the velocity at which a particular sized drop or jet causes strength loss in the target. With single impact testing, the threshold velocity can be determined by obtaining 'residual' strength curves, i.e measuring the specimen strength after impact, and recording the velocity at which strength loss takes place. This is much more accurate than recording the onset of visual damage (by eye or low power microscopy) since the defects controlling the strength may still be submicroscopic (see, for example, Field *et al.* 1979; van der Zwaag & Field 1983*b*). With MIJA a different approach is used and the number of impacts to cause visible damage at a particular impact velocity is recorded (see later).

(*f*) *Frictional properties*

Friction measurements were made with an apparatus of the reciprocating type. A diamond stylus moves to and fro along a 1 mm track with a chosen load. Data are recorded at 10 μm separations, though the readings from the ends of the track where the stylus stops and reverses are not used. To avoid having to store too much data the computer is programmed to only record data when significant changes of friction coefficient occur. The programme allows the operator to abort or pause the experiment or to manually override the recorded algorithm and force the apparatus to take readings. For a full discussion, see Hayward & Field (1988).

4. Results

(*a*) *Hardness and modulus*

The Knoop hardness of natural diamond falls in the range 55–113 GPa depending on the crystal orientation and whether the diamond is type I or type II (see chapters by Brookes in Field 1979, 1992). The extreme hardness of diamond makes its measurement difficult and skill is required if the indenter is not to be broken. The fact that CVD diamond can also damage indenters suggests that it can also be hard. A selection of our and other people's data are given in table 1. Beetz *et al.* (1990) found that films deposited at lower methane concentration, 0.11% CH_4 in H_2, had large crystallite sizes of 5–8 μm and an average hardness and modulus of 31 and 541 GPa respectively. A higher CH_4 concentration of 0.99% in H_2 resulted in finer crystallites of *ca.* 0.5 μm and average hardness and modulus of 65 and 875 GPa respectively. According to Beetz *et al.* there is more Sp^3 bonding in the lower hardness film which is counter intuitive but changes in crystal size may be the dominant feature. (It is regrettable that Beetz *et al.* compromise the value of their results by so many discrepancies between data in their figures and statements in the text.) Similar values of hardness and modulus were found for CVD films by Tsukamoto *et al.* (1987). Our own work and that of many authors have moduli values for CVD diamond in the range 700–900 GPa. Diamond-like carbon (DLC) films, as expected, tend to have lower hardness and modulus values (though note that a hardness of 20 GPa is still about three times that of tool steel!). The very high values (see table 1), close to the aggregate bulk value for diamond of 1041 GPa (Ruoff 1979) are surprising. In our experience, the bulge test can give high modulus values if the film has corrugations and this may be an important factor. The accuracy of the values obtained in the vibrating reed and bulge tests is not limited at present by the intrinsic accuracy of the tests (to a few percent), but by the imperfections of the films; non-uniformities in thickness, residual stresses, etc.

(*b*) *Erosion and strength*

(i) *Solid particle erosion*

The erosion resistances of bulk diamond and PCD composites are the highest yet recorded (Hayward & Field 1990; Feng & Field 1990; Vaughan & Ball 1991). With sand as the erodent, there is negligible erosion until velocities in excess of *ca.* 100 m s^{-1} and then the steady state erosion is at a rate of *ca.* 0.05 mg per kg of erodent. Bulk diamond first forms a network of cracks and material loss only starts when cracks intersect. Polycrystalline diamond composites erode primarily by loss of the binder metal (usually cobalt).

To date, four CVD films have been studied (Feng *et al.* 1992*a*); (*a*) polished film (with a surface roughness of 8 nm CLA and a film thickness of *ca.* 15 μm), deposited on a silicon nitride substrate by microwave plasma CVD; (*b*) unpolished film (with a surface roughness of 1 μm CLA and a film thickness of *ca.* 6 μm), deposited on silicon by hot filament CVD with (100) facets dominating; (*c*) unpolished film (the preparation method, surface roughness, thickness and substrate were the same as (*b*), but diamond grains with (111) facets dominating); and (*d*) unpolished free-standing film deposited by hot filament CVD with a grain size of 2–4 μm and a film thickness of 25 μm. Films (*a*), (*b*) and (*c*) were used in the solid impact experiments and film (*d*) for fractographic study following bending. Film (*a*) was also used in static indentation experiments to determine fracture strength and to make a comparison with results obtained from impact experiments. Detailed information about the preparations and characterizations of the CVD diamond films can be found in Srivinyonon *et al.* (1991) and Nishimura (1991).

In these erosion experiments impact velocity and exposure time were varied and the conditions for film cracking and delamination obtained. The first experiment was performed at an impact velocity of 59 m s^{-1}, with a flux rate of 4.6 kg m^{-2} s^{-1} and an

Table 1. *Thin film properties*

(HFA, hot filament assisted; MPA, microplasma assisted; P, plasma; DLC, diamond-like carbon; Si, silicon substrate.)

film type	hardness GPa	modulus GPa	measurement method	reference
MPACVD/Si (3–11 μm)		740 ± 20 (3.7 μm) 710 ± 20 (5.5 μm)	vibrating reed	this paper
MPACVD free standing				
(230–245 μm)		745 ± 10	vibrating reed	this paper
(150–180 μm)		780 ± 10		
MPACVD		680–790	vibrating membrane	Berry *et al.* (1990)
HFACVD/Si	31	540[a]	nano indentation	Beetz *et al.* (1990)
35 μm	65	875[b]		
MPACVD				
9.6 μm		864, 893[c]	bulge test	Cardinale & Tustison (1990, 1991)
16 μm		1054, 1095[c]		
DLC/Si 0.25–0.4 μm	20–48	480–850	H = nano indentation for E = beam bending, with both end supported	Hoshino *et al.* (1989)
P CVD DLC/Si 0.1–0.4 μm	63–67	810–820	H = nano indentation E = beam bending, with ends supported on fulcrums	Tsukamoto *et al.* (1987)

[a] Low % CH_4. [b] High % CH_4. [c] ν taken as 0.1 and 0.07 respectively.

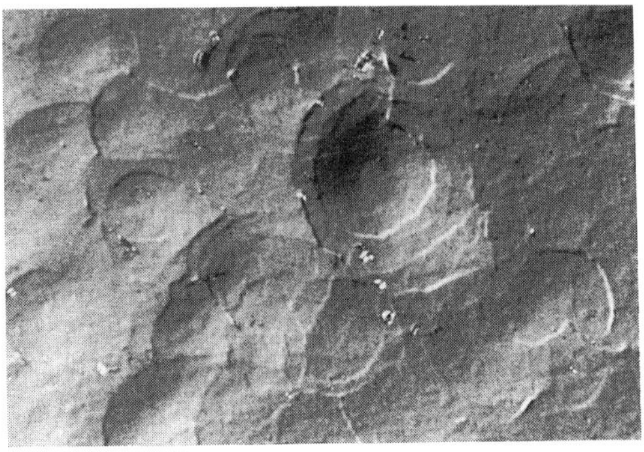

Figure 4. Optical micrographs of an eroded CVD diamond film at an impact velocity of $34\ m\ s^{-1}$ for 10 s.

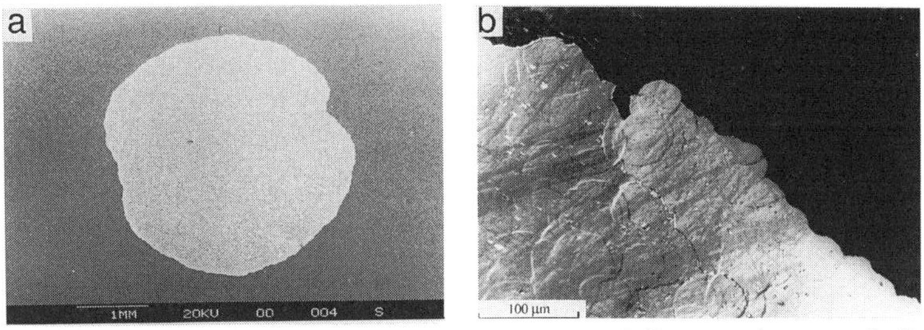

Figure 5. (*a*) Optical micrograph of an eroded CVD diamond film at an impact velocity of $59\ m\ s^{-1}$ for 70 s. (*b*) SEM micrograph of an area near the broken edge.

exposure time of 10 s, and ring cracks were found on the surfaces after impact (see figure 4). The experiment was then carried out at a lower sand velocity, $34\ m\ s^{-1}$, with the same flux rate and for the same time of exposure, and again ring cracks were formed on the surfaces but with a lower crack density. There was no significant difference in the sizes of ring cracks observed at these two velocities. No diamond film removal occurred at this stage for either impact velocity. Experiments were then performed at an impact velocity of $59\ m\ s^{-1}$ for an increased time of impact. It was found that film removal from the substrate started after an exposure time of *ca.* 70 s. Figure 5*a* shows an SEM micrograph of the area where the film was delaminated. Figure 5*b* is an optical micrograph from the edge of the impacted area and shows that the boundary is defined by the ring cracks. The film surface adjacent to a delaminated area can be used to study what happens just before film removal starts because of the lower impact flux in this area owing to the shadowing effect of the rubber mask. Examination by optical microscopy showed that the film surface became increasingly wavy compared with other areas. This is presumably due to partial debonding of the film from the substrate. The fractured edge tends to curve out of the substrate surface because of residual stresses in the film.

The damage process for a CVD diamond film by sand particle impact follows the sequence (i) formation of ring cracks, (ii) debonding and/or penetration of the film, and (iii) removal or delamination of the film from the substrate. The formation of

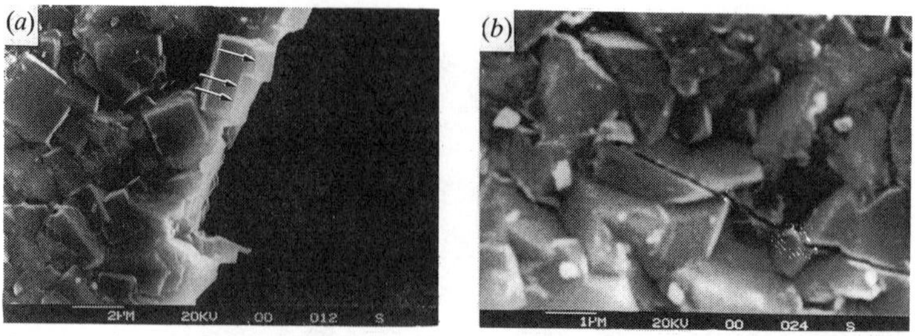

Figure 6. SEM micrographs of fractured surfaces produced by sand particle impact (*a*) on an unpolished (100) facet dominated film, (*b*) on an unpolished (111) facet dominated film. Both contain examples of transgranular fracture.

ring cracks at the initial stage is a typical fracture phenomenon for brittle materials. This was also observed in the impact of natural diamond by sand particles using the same apparatus (Feng & Field 1990). However, the material removal mechanisms are very different: in the case of CVD diamond, the premature removal of the film is caused by the delamination; while in the case of natural diamonds it occurs only when ring or cone cracks intersect to separate small volumes of diamond from the bulk.

(ii) *Fracture morphology and strength*

Scanning electron microscopy of the fracture surfaces of the eroded films and also CVD free-standing films showed many examples of transgranular fracture. Figure 6*a* shows a fractured edge on a (100) facet dominated film: the arrowed region clearly indicates transgranular fracture. Figure 6*b* shows a similar micrograph for a (111) facet dominated film. Transgranular fracture was also common when free-standing films were broken by bending.

The fracture strengths of the films were estimated (see Feng *et al.* (1992*a*) for full details) by measuring the ring crack diameters on the eroded specimens and assuming that they were formed by spherical sand particles of diameters in the range 300–600 μm. Additionally, ring cracks were formed by indenting with a 0.39 mm tungsten carbide sphere. In both cases, Hertz theory (elastic sphere indenting on elastic half-space) was used.

Both theory (El-Sherbiney & Halling 1976; van der Zwaag & Field 1982) and experiment show that for a hard coating with a Young's modulus about three times that of the substrate and a coating thickness of $d/a_0 \approx 0.5$ (where d is the coating thickness and a_0 the contact radius calculated by Hertz theory), the measured contact size on the coating is almost the same as that given by Hertz theory for a half-space of the coating material. The result of these measurements is that ring cracks form on CVD film at a critical contact pressure (load divided by crack area) of 3.2 GPa compared with a critical contact pressure of 10–13.5 GPa for natural diamond (data from Howes 1962). Assuming the ring crack propagates at 90° to the surface and that the Poisson ratio is 0.07 then Hertz theory predicts a tensile strength for CVD diamond of 1.4 GPa and 4–6 GPa for bulk natural diamonds. (Note Feng *et al.* (1992*a*) missed a factor 2 which gave their values twice as high as these.) Windischmann *et al.* (1991) recorded CVD strengths in the range 1–5 GPa using a bulge test.

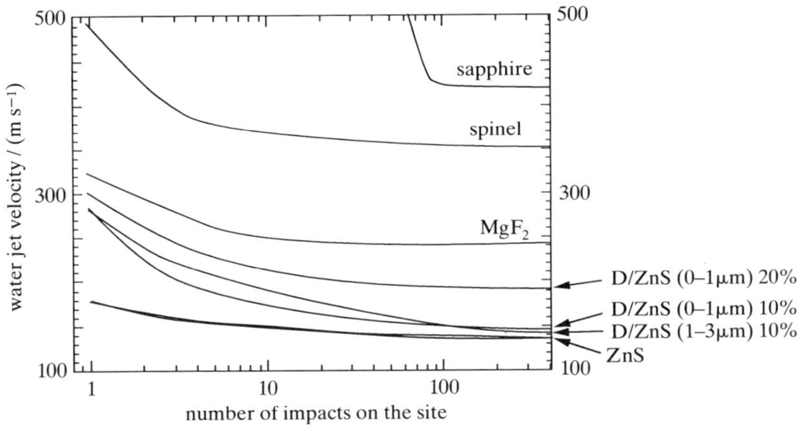

Figure 7. Liquid impact erosion data for a range of materials. A curve separates damaged and undamaged sites. The velocities after 400 impacts are taken as 'threshold' values. Data for a 0.8 mm nozzle.

Table 2. *Threshold data for liquid impact*
(MITV, multiple impact threshold velocity.)

material	MITV 0.8 mm jet/(m s^{-1})	MITV 2 mm drop/(m s^{-1})
zinc sulphide	130	190
magnesium fluoride	205	290
spinel	350	460
sapphire (25 mm disc)	420	*ca.* 535
natural diamond	*ca.* 530	*ca.* 600
amorphic diamond on ZnS (1 μm)	*ca.* 160	*ca.* 235
ZnS/D 20%, 0–1 μm	190	270
ZnS/D 10%, 1–3 μm	140	205
ZnS/D 10%, 0–1 μm	140	205
CVD diamond	< 200	< 285

(iii) *Liquid impact erosion*

Data on the liquid impact performance of a range of infrared transmitting materials are given in figure 7 and table 2. The MIJA was programmed to impact a succession of sites at different, pre-selected velocities (17 in the data illustrated in figure 7). The specimen, still on its computer-controlled stage was then moved for microscopic examination. The sites at which damage was visible were recorded. The procedure was then repeated for larger numbers of impacts with impacts on the original sites at the same pre-selected velocity. Microscopic examination was made regularly until undamaged sites had received 400 impacts. After this number of impacts it has been shown that sites where the 'threshold' velocity has been exceeded exhibit visual damage. Sites which do not show damage at this stage are below the critical threshold velocity.

Zinc sulphide, ZnS, is an important infrared window material in the 8–12 μm range but it is a relatively weak material with a low threshold for damage. It is interesting that ZnS strengthened by diamond particles has an improved erosion

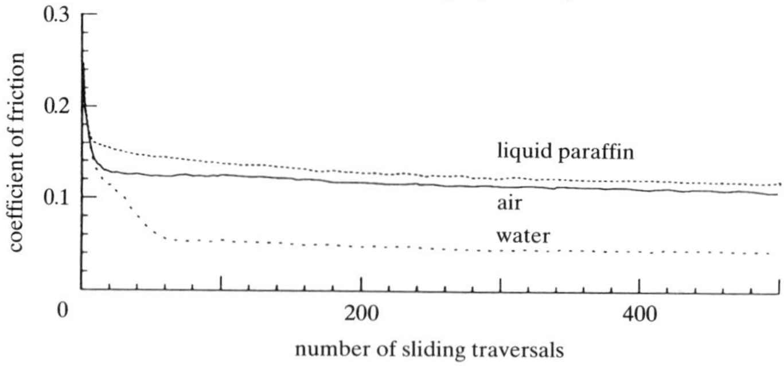

Figure 8. Frictional coefficient of diamond sliding on a CVD diamond coating against number of sliding traversals in air, liquid, paraffin and water.

response. (For details of how these composites are fabricated and their strength and optical properties see Xue *et al.* (1990) and Farquhar *et al.* (1990).)

Sapphire, germanium (Ge), magnesium fluoride and spinel have superior thresholds to ZnS but Ge has poor transmission above 370 K and the others are of prime interest in the 3–5 μm window range.

Rain erosion threshold velocity data on natural diamond are, not surprisingly, scarce. To date we have tested one natural diamond; a 6 mm diameter, 2 mm thick type IIa diamond. The threshold figure for impact by a 0.8 mm jet (equivalent to a 4 mm drop) was *ca.* 530 m s^{-1}. The equivalent 2 mm drop threshold, calculated by equations given in Hand *et al.* (1991), is *ca.* 600 m s^{-1}. This may be artificially low because of the small specimen size which would give sizeable reflected tensile pulses. In other experiments, amorphic diamond CVD coatings on ZnS detached at a velocity of *ca.* 160 m s^{-1} (2 mm drop equivalent of *ca.* 235 m s^{-1}). Experiments on a bulk CVD layer of 300 μm thickness had a threshold velocity of *ca.* 200 m s^{-1} (2 mm drop equivalent of *ca.* 285 m s^{-1}). However, it is likely that this specimen was considerably weakened by the presence of microcracks before testing, so this value is probably much lower than what could be achieved with a carefully selected specimen.

(iv) *Frictional properties*

The friction of natural diamond has been of considerable interest for many years due to its low value. The precise figure depends on the crystal force, the sliding direction, the material it slides against and the load. For diamond sliding on diamond, in air, values of 0.05 to 0.1 are common and in the presence of water can be even lower (for a recent review see Tabor & Field (1992)).

A potential advantage of CVD diamond is that the friction should be more isotropic with sliding direction. Research to date on CVD diamond has shown the following. (i) The friction coefficient is usually two or three times that of natural diamond for diamond sliding on diamond. Feng & Field (1992) have, for example, recorded values of *ca.* 0.25 for the friction coefficient of CVD diamond coated on silicon. This friction, for a small number of traversals, remained constant up to a load of 2N (contact pressure *ca.* 10 GPa) when coating delamination occurred. (ii) With large numbers of traversals, the coefficient of friction decreases (see figure 8). The higher initial value has been attributed to the greater surface roughness of the CVD diamond (Feng & Field 1991; Hayward *et al.* 1992). As the sliding progresses, the friction track becomes smoother. (iii) The friction response to different oils (small effect) and to the pressure of water (large effect; see figure 8) is qualitatively the same as natural diamond (Feng & Field 1991). (iv) There is some evidence that at heavy loads the CVD layer is plastically deformed (Feng & Field 1991).

5. Discussion

The measurement of CVD diamond properties is at an early stage, but is essential if CVD coatings and self-supporting layers are to be optimized. Various techniques are available for modulus measurements, but as noted earlier, their value is limited less by the accuracy of the techniques than by problems with the films themselves. The measurements will become more meaningful as CVD layers become thicker, more uniform and less stressed.

We have recorded tensile strengths for CVD layers of *ca.* 1 GPa and this agrees with values by other workers. The value is less than for good quality diamond by a factor of 2–5 times. Field (1979) has estimated that sharp-ended microcracks of *ca.* 0.5 μm could explain the tensile strength of natural diamond. This would suggest flaws of *ca.* 2–10 μm in CVD diamond. The observation that *transgranular* fracture is common is encouraging since it suggests that the grain boundaries in CVD diamond can be strong.

The erosion and friction studies show that delamination can be a major problem. The potential of free-standing layers for infrared windows is exciting, since natural diamond has a very high erosion resistance against solid particles and a high threshold for damage for liquid drops. In the liquid impact test the threshold of natural diamond was probably artificially low because the small specimen size gives stress wave reinforcements from reflected waves. CVD layers would have similar thicknesses to natural diamond specimens and much greater lateral dimensions which would reduce this problem.

This work has been supported by grants from MOD Proc. Executive (now DRA) and De Beers Industrial Diamond Division. We thank De Beers, the Naval Research Laboratory, Washington, D.C., U.S.A., Professor Y. Tzeng, Auburn University, U.S.A. and Professor R. Raj, Cornell, U.S.A., for the provision of samples.

References

Allen, M. G., Mehregany, M., Howe, R. T. & Senturia, S. D. 1987 *Appl. Phys. Lett.* **51**, 241–243.

Andrews, D. R., Walley, S. M. & Field, J. E. 1983 In *Proc. 6th Int. Conf. on Erosion by liquid and solid impact* (ed. J. E. Field), paper 36. Cambridge: Cavendish Laboratory.

Anthony, T. R., Banholzer, W. F., Fleischer, J. F., Wei, L., Kuo, P. K., Thomas, R. L. & Pryor, R. W. 1990 *Phys. Rev.* B **42**, 1104–1111.

Beetz, C. P., Cooper, C. V., Perry, T. A. 1990 *J. Mater. Res.* **5**, 2555–2561.

Berry, B. S., Pritchet, W. C., Cuomo, J. J., Guarnieri, C. R. & Whitehair, S. J. 1990 *Appl. Phys. Lett.* **57**, 302–303.

Bowden, F. P. & Brunton, J. H. 1961 *Proc. R. Soc. Lond.* A **263**, 433–450.

Bowden, F. P. & Field, J. E. 1964 *Proc. R. Soc. Lond.* A **282**, 331–352.

Bromley, E. I., Randall, J. N., Flanders, D. C. & Mountain, R. W. 1983 *J. Vac. Sci. Technol.* B **1**, 1364–1366.

Brookes, C. A. 1979 In *The properties of diamond* (ed. J. E. Field), pp. 383–402. London: Academic Press.

Brookes, C. A. 1992 In *The properties of natural and synthetic diamond* (ed. J. E. Field), pp. 515–546. London: Academic Press.

Cardinale, G. F. & Tustison, R. W. 1990 *SPIE*, 1325, Diamond Optics III, 90–98.

Cardinale, G. F. & Tustison, R. W. 1991 *J. Vac. Sci. Technol.* A **9**, 2204–2208.

Davidson, J. L., Ramesham, R. & Ellis, C. 1990 *J. electrochem. Soc.* **137**, 3203–3205.

El-Sherbiney, M. G. D. & Halling, J. 1976 *Wear* **40**, 325–337.

Farquhar, D. S., Raj, R., Phoenix, S. L. 1990 *J. Am. Ceram. Soc.* **73**, 3074–3080.

Feng, Z. & Field, J. E. 1990 *J. Hard. Mater.* **1**, 273–287.

Feng, Z. & Field, J. E. 1991 *Surf. Coatings Technol.* **47**, 631–645.

Feng, Z. & Field, J. E. 1992 *J. Phys.* D **25**, A33–A37.

Feng, Z., Tzeng, Y. & Field, J. E. 1992*a* *Thin Solid Films* **212**, 35–42.

Feng, Z., Tzeng, Y. & Field, J. E. 1992*b* *J. Phys.* D **25**, 1418–1424.

Field, J. E. 1979 *The properties of diamond* (ed. J. E. Field). London: Academic Press.

Field, J. E. 1992 *The properties of natural and synthetic diamond* (ed. J. E. Field). London: Academic Press.

Field, J. E., Gorham, D. A., Hagan, J. T., Matthewson, M. J., Swain, M. V. & van der Zwaag, S. 1979 In *Proc. 5th Int. Conf. on Erosion by Liquid and Solid Impact* (ed. J. E. Field), paper 13. Cambridge: Cavendish Laboratory.

Grimsditch, M. H. & Ramdas, A. K. 1975 *Phys. Rev.* B **11**, 3139–3148.

Hand, R. J., Field, J. E. & Townsend, D. 1991 *J. appl. Phys.* **70**, 7111–7118.

Hayward, I. P. & Field, J. E. 1988 *J. Phys.* E **21**, 753–756.

Hayward, I. P. & Field, J. E. 1990 *J. Hard. Mater.* **1**, 53–64.

Hayward, I. P., Singer, I. L. & Seitzmann, L. E. 1992 *Wear* **157**, 215–228.

Hoshino, S., Fujini, K., Shohata, N., Yamaguchi, H., Tsukamoto, Y. & Yanagisawa, M. 1989 *J. appl. Phys.* **65**, 1918–1922.

Howes, V. R. 1962 *Proc. Phys. Soc.* **80**, 78–80.

Isukamoto, Y., Yamaguchi, H. & Yanagisawa, M. 1987 *Thin Solid Films* **154**, 171–181.

Kinbara, A., Baba, S., Matuda, N. & Takamisawa, K. 1981 *Thin Solid Films* **84**, 205–212.

Lesser, M. B. 1981 *Proc. R. Soc. Lond.* A **377**, 289–308.

Lesser, M. B. & Field, J. E. 1983 *A. Rev. Fluid. Mech.* **15**, 97–122.

McHargue, C. J. 1990 *Proc. NATO Adv. Study. Inst.* Italy, Plenum Press.

Nishimura, K. 1991 *Diamond Optics IV SPIE* **1534**, 199–205.

Oliver, W. C. & McHargue, C. J. 1988 *Thin Solid Films* **161**, 117–122.

Ono, A., Baba, T., Funamoto, H. & Nishihawa, A. 1986 *Jap. J. appl. Phys.* **25**, L808–810.

Pethica, J. B., Hutchings, R. & Oliver, W. C. 1983 *Phil. Mag.* A **48**, 593–606.

Pryor, R. W., Kuo, P. K., Wei, L. & Thomas, R. L. 1990 *Rev. Prog. Quant. NDE* **9**, 1123–1128.

Ruoff, A. L. 1979 In *High pressure science and technology* (ed. K. D. Timmerhaus & M. S. Barker), vol. 2, pp. 525–548. Plenum.

Seward, C. R., Pickles, C. S. J. & Field, J. E. 1990 In *Proc. SPIE Conf.* on *Window and Dome Technologies and Materials*, vol. 1326, pp. 280–290.

Smith, J., Holiday, P., Dehbi-Alaoui, A. & Mathews, A. 1992 *Diamond Related Mater.* **1**, 355–359.

Srivinyonon, T., Philips, R., Cutshaw, C., Joseph, A. J. & Tzeng, Y. 1991 In *Proc. 2nd Int. Conf. on New Diamond Science and Technology*, pp. 581–586. Pittsburgh: Materials Research Society. PA.

Tabor, D. & Field, J. E. 1992 In *The properties of natural and synthetic diamond* (ed. J. E. Field), pp. 547–571. London: Academic Press.

Tsukamoto, Y., Yamaguchi, H. & Yanagisawa, M. 1987 *Thin Solid Films* **154**, 171–181.

van der Zwaag, S. & Field, J. E. 1982 *Phil. Mag.* A **46**, 133–150.

van der Zwaag, S. & Field, J. E. 1983*a* *Phil. Mag.* A **48**, 767–777.

van der Zwaag, S. & Field, J. E. 1983*b* *Engng Frac. Mech.* **17**, 367–379.

van der Zwaag, S., Dear, J. P. & Field, J. E. 1986 *Phil. Mag.* A **53**, 101–111.

Vaugnan, R. A. & Ball, A. 1991 *J. Hard. Mater.* **2**, 257–269.

Walley, S. M. & Field, J. E. 1987 *Phil. Trans. R. Soc. Lond.* A **321**, 277–303.

Windischmann, H., Epps, G. F. & Ceasar, G. P. 1991 In *Proc. 2nd Int Conf. on New Diamond Science and Technology*, pp. 762–772. Pittsburgh: Materials Research Society.

Xue, L. A., Farquhar, D. S., Noh, T. W., Sievers, A. J. & Raj, R. 1990 *Acta metall. Mater.* **38**, 1743–1752.

Yang, W. M. C., Tsakalakos, T. & Hilliard, J. E. 1977 *J. appl. Phys.* **48**, 876–879.

Discussion

L. M. Brown (*Cavendish Laboratory, Cambridge, U.K.*): It is an interesting observation that polycrystalline diamond films show largely transgranular fracture. Two influences may be at work here. First, if there is an intergranular layer of amorphous carbon it may provide some local plasticity and therefore a higher work of fracture for intergranular cracks than for transgranular ones, which will experience little or no plastic blunting. Secondly, I have the impression that in many films the diamond grains experience large internal stresses. Certainly, the multiply-twinned grains will have a strong circumferential tension at the external surfaces, unless plastic relief of stresses has occurred. The grains should be individually very fragile. Perhaps these two factors operate to explain the observed mode of fracture.

J. E. Field. Certainly it is encouraging that the grain boundaries are not major sources of weakness.

J. E. Butler (*Naval Research Laboratory, Washington, D.C., U.S.A.*): (*a*) Has the coefficient of friction been measured using other oxygen containing lubricants, e.g. alcohols, ketones, etc.? (*b*) Please comment on the role of tribochemical polishing of rough surface asperities by oxygen species on the lowering of the fraction coefficient? (*c*) Please comment on the degree to which the bulk mechanical properties of polycrystalline CVD diamond is controlled by the nature of the grain boundaries versus the defects in the individual grains?

J. E. Field. To date we have studied conventional lubricants and water, which, as shown in the paper, has a dramatic effect in reducing friction. At present, we are doing experiments with water at different pH values. I agree it would be useful to study oxygen-containing lubricants. It is difficult to comment briefly on the effect of surface roughness on diamond friction since there is still controversy about the mechanisms which operate in diamond friction (see Tabor & Field 1992). However, research at Cambridge and also at your laboratory has shown clearly that the coefficient of friction reduces as the diamond surface becomes smooth. My answer to Professor Brown gives our current thinking on question (*c*).

9

Deposition of diamond-like carbon

BY J. ROBERTSON

Diamond-like carbon refers to forms of amorphous carbon and hydrogenated amorphous carbon containing a sizeable fraction of sp^3 bonding, which makes them mechanically hard, infrared transparent and chemically inert. This paper discusses the various thin film deposition processes used to form diamond-like carbon and the deposition mechanisms responsible for promoting the metastable sp^3 bonding.

1. Introduction

Crystalline carbon forms a number of allotropes because the C atom is able to exert both sp^3 and sp^2 hybridization. Graphite, the stable allotrope, is a layered, sp^2-bonded metal. Diamond, only 0.02 eV less stable at 0 K, is a sp^3 bonded, wide band-gap semiconductor with the highest bulk modulus, hardness and atom number density of any solid.

There are also many non-crystalline forms of carbon. The familiar non-crystalline carbons such as soot, evaporated amorphous carbon, or glassy carbon are sp^2 bonded. There is presently great interest in those types of amorphous carbon (a-C) and hydrogenated amorphous carbon (a-C:H) containing significant sp^3 bonding (Angus & Hayman 1988; Robertson 1986, 1991*a*, *b*, 1992*a*). Their sp^3 bonding confer on this 'diamond-like carbon' (DLC) many of the valuable properties of diamond itself, such as high mechanical hardness, low friction, transparency and chemical inertness. In general, DLCs contain a mixture of both sp^3 and sp^2 sites and hydrogen (table 1). They can therefore be considered to be intermediate between diamond, graphite and a polymeric hydrocarbon. This mixed bonding means that their properties are generally poorer than those of diamond. Nevertheless, their low cost, smoothness, low deposition temperature and ability to be deposited on ferrous substrates makes DLC a competitive coating material which is available now. The drawbacks of DLC coatings are their poor thermal stability and high internal stress.

2. Electronic structure

The electronic structure of a-C and a-C:H was reviewed by Robertson (1986, 1991*a*, *b*). Amorphous carbons contain both sp^3 and sp^2 sites. The sp^3 sites form σ bonds while the sp^2 sites form both σ and π bonds. The strong σ bonds form the skeleton of the covalent network. The π bonds favour the segregation of sp^2 sites into graphitic clusters, embedded in a sp^3 bonded matrix. The π states lie closest to the Fermi Level, so they control the electronic properties like the band gap (figure 1). The band gap varies with cluster size as

$$E_g = 6/M^{\frac{1}{2}}\,\mathrm{eV}, \tag{1}$$

where M is the number of six-fold rings in the cluster (Robertson & O'Reilly 1987).

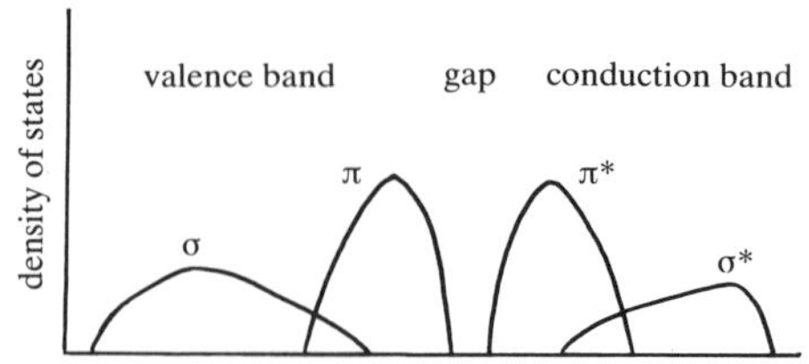

Figure 1. Schematic electronic density of states of amorphous carbons.

Table 1. *Properties of various forms of carbon*

	density	hardness/GPa	sp^3 (%)	H (at. %)	band gap/eV
diamond	3.515	100	100	0	5.5
graphite	2.267	—	0	0	−0.04
glassy C	1.3–1.55	2–3	0	—	0.01
evap a-C	1.8–2.0	2–5	< 5	—	0.4–0.7
MSIB a-C	2.5–3.0	100	90 ± 5	*ca.* 9	1.5–2.5
sput a-C	2.0–2.4	10–14	*ca.* 95	—	*ca.* 0.5
laser nc-C	—	40	—	—	—
hard a-C:H	1.6–2.2	10–40	30–60	20–40	0.8–1.5
soft a-C:H	0.9–1.6	< 5	50–80	40–65	1.5–4
polythene	0.92	0.01	100	67	6

This suggests a two-phase model of amorphous carbons. This model works best for a-C:H, where the sp^3-bonded phase can be a highly cross-linked network as a 'hard' a-C:H, or an open polymeric network as in 'soft' a-C:H. The relatively small band gap, the Raman spectra and the luminescence of a-C:H all support this model (Robertson 1992*b*). The electronic spectra have been studied experimentally by photoemission (Oelhafen *et al.* 1991) and electron energy loss (Fink *et al.* 1983).

3. Mechanical properties

The high hardness, wear resistance and low friction are the most useful valuable properties of DLC. Robertson (1991*b*, 1992*c*, *d*) showed how each of these properties are fundamentally related to Young's modulus E, and how E depends on the underlying bonding. The high modulus of diamond results from its strong, directional bonds. The moduli of DLCs are lower because of their finite sp^2 and H content. E is found to vary with mean C–C coordination of the a-C(:H) network, r, as

$$E = E_0 c_{sp^3}\{(r-2.4)/(r_0-2.4)\}, \tag{2}$$

where E_0 is the modulus of diamond at $r_0 = 4$, and c_{sp^3} is the fraction of sp^3 bonding. The sp^2 fraction of the network tends to contribute no modulus because of its locally layered bonding. Thus, in the two phase model, Young's modulus depends only on the sp^3 phase, its concentration and its mean C–C coordination. This model has been found to describe well the modulus of a-C:H, as seen in figure 2. It is less successful for sputtered a-C, whose sizeable modulus of 11–14 GPa (Cho *et al.* 1990) but small sp^3 fraction (*ca.* 5%) implies that there is considerable sp^2-type interlayer cross-linking. The linking may somewhat resemble that in the recently proposed C sponge structures (O'Keeffe *et al.* 1992).

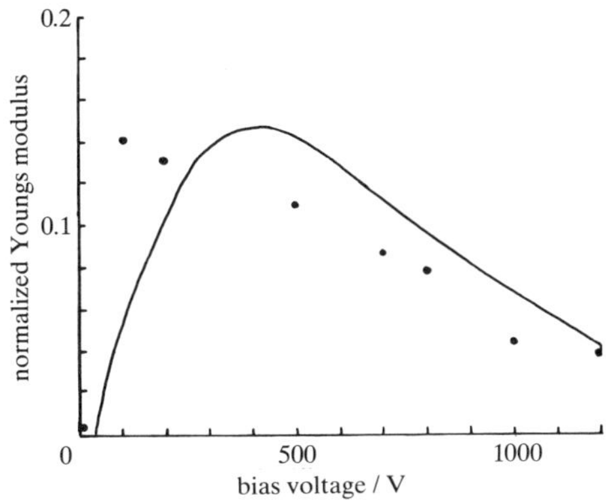

Figure 2. Comparison of experimental (Jiang 1989) and calculated Young's modulus of plasma-deposited a-C:H as a function of bias voltage.

4. Deposition methods

Diamond-like carbon can be prepared by a wide variety of deposition methods, such as ion beam deposition, mass-selected ion beam deposition, ion beam sputtering, magnetron sputtering, plasma deposition and laser plasma deposition. A common feature of each method is the exposure of the growing film to bombardment by ions of medium energy, 20–500 V, which appears necessary to promote sp^3 bonding.

In ion beam deposition, carbon ions are generated by sputtering carbon electrodes by an Ar plasma and then accelerated towards the substrate by a bias electrode. Higher growth rates are possible by generating ions from a hydrocarbon source gas. A solid carbon source gives ions of the form C_m^+, while a hydrocarbon source gives $C_mH_n^+$ ions. In both cases the substrate also receives a large flux of neutral species such as unionized Ar or source gas.

Deposition of a single ion species is possible if the ion beam is passed through a magnetic mass analyser for e/m selection. The analyser filters neutrals, cluster species, graphitic fragments and impurities from the beam and allows only a pure beam of C^+ (or C^-) ions to reach the substrate. This mass-selected ion beam (MSIB) method was first used by Aksenov *et al.* (1970). Structural studies indicate a form of a-C with the highest fraction of sp^3 bonding of those from any present deposition process. The deposition rate for this method can be maximized by using a carbon arc as an ion source. The arc is confined magnetically for stability. The main practical problem with this method is the high compressive stress in the films, which limits adhesion for films over 500 Å† thick (McKenzie *et al.* 1991*a*).

Various sputtering methods can be used. In ion beam sputtering, a beam of typically 1 kV Ar ions is directed at a graphite target. An angle of incidence of 30–45° is used to maximize the yield. A second Ar ion beam can be directed at the substrate to provide bombardment of the growing film. Higher deposition rates can be achieved by magnetron sputtering (Savvides 1990). Here an Ar plasma is used to both sputter from the target and bombard the growing film. Deposition rates are typically 3 Å s^{-1} and increase linearly with RF power. Ion energies are of order 20 eV

† 1 Å $= 10^{-10}$ m $= 10^{-1}$ nm.

and decline slowly with increasing power or gas pressure. The ion/neutral flux ratio falls with increasing sputtering power and gas pressure. The advantage of sputtering is good process control and ease of scale-up to larger apparatus.

One of the most popular methods is RF plasma-deposition (PD) from a hydrocarbon source gas (Koidl *et al.* 1990). RF power is capacitively coupled to the substrate electrode and the counter electrode is either a second electrode or just the grounded walls of the deposition chamber. If the RF frequency is greater than the ion plasma frequency, *ca.* 2–5 MHz, the electrons can follow the RF voltage but the ions cannot. The powered electrode acquires a negative self-bias because of the large difference in electrode size and also in the electron and ion mobilities. The DC bias is largely dropped across an ion sheath in front of the cathode, and it accelerates the ions towards the cathode. The bias voltage V_b varies with RF power W and operating pressure P as $V_b = k(W/P)^{\frac{1}{2}}$. The ion energy E_i depends on V_b and the ion mean free path in the sheath. At low pressures in the absence of collisions $E_i \approx V_b$, while at higher pressures there is a spectrum of ion energies with a mean value of $E_i \approx k' V_b P^{-\frac{1}{2}}$, or $E_b \approx 0.6\ V_b$ for typical pressures of 3 Pa (Koidl *et al.* 1990). The deposition rate for a given source gas tends to vary linearly with bias voltage and gas pressure (Koidl *et al.* 1990; Zou *et al.* 1989, 1990). The rate is highest for gases of low ionization potentials and large molecular weights. Films deposited from acetylene appear to have the best properties, having the highest hardness and a reasonable deposition rate.

A carbon ion plasma can also be produced by the laser ablation of a graphite target (Davanloo *et al.* 1992). The plasma somewhat resembles that formed by a cathodic arc. The resulting film is diamond-like if the laser power exceeds a certain threshold. The DLC is found to consist of nano-scale mixture of diamond grains embedded in an a-C matrix. It has the advantage of high hardness yet moderate internal stress.

The preferred deposition method depends on whether the DLC preparation is for practical application or investigative studies. For practical applications, factors such as deposition rate, film properties, film adherence, sample coverage, process control and scale up are of importance, which favour sputtering, plasma-deposition and ion-plating. MSIB and PD methods are important in fundamental studies.

5. Deposition mechanisms

(*a*) MSIB a-C

A variety of structural measurements (electron energy loss, diffraction) show that MSIB a-C has a high degree (up to 95%) of sp^3 bonding and little hydrogen (McKenzie *et al.* 1991*a*, *b*; Gaskell *et al.* 1991). Figure 3 shows the variation of density with C^+ ion energy, after McKenzie *et al.* (1991*a*). Density is expected to vary closely with sp^3 fraction because of the large density difference between graphite and diamond. The sp^3 fraction increases from 20 eV to reach a peak at about 40 eV before gradually declining at higher energies. Koskinen (1898) and Ishikawa *et al.* (1987) found that other properties such as hardness peak at the same ion energy as density (although they found a higher optimum energy).

The mechanism that promotes sp^3 bonding in a-C is still contentious. Spencer *et al.* (1976) suggested a preferential sputtering of sp^2 sites but Lifshitz *et al.* (1990) noted that this was unlikely due to the low sputtering yield of carbon. Lifshitz *et al.* (1990) proposed a 'subplantation' mechanism in which a preferential displacement of sp^2 sites led to an accumulation of sp^3 sites. This idea was based on estimates of

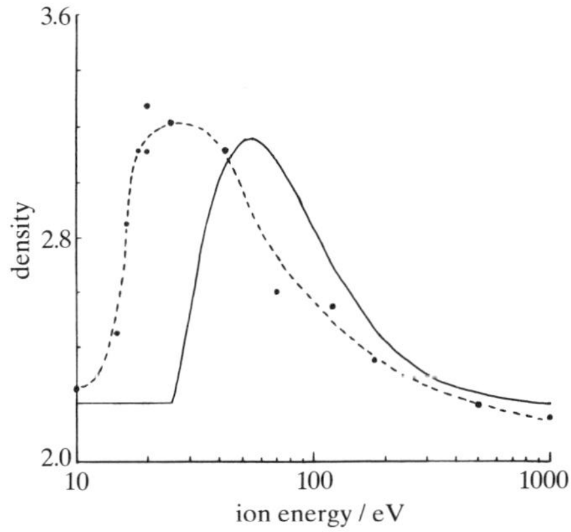

Figure 3. Comparison of experimental (McKenzie *et al.* 1991*b*) and calculated variation of density with ion energy of MSIB a-C. ---, Experimental; ——, calculated.

the displacement threshold of graphite and diamond as 30 eV and 80 eV respectively. However, recent measurements have found quite similar thresholds for graphite (35 eV, Steffen *et al.* (1992)) and diamond (37–47 eV, Koike *et al.* (1992)), which invalidates this mechanism. McKenzie *et al.* (1991) proposed that the high intrinsic stress moved MSIB a-C into the stability domain of diamond and thereby stabilized sp^3 bonding. Molecular dynamics simulations support a subplantation mechanism (Pailthorpe 1991; Kaukonen & Nieminen 1991).

It is necessary to analyse deposition further in terms of elementary processes to understand the deposition mechanism. I propose that sp^3 bonding is promoted by the ion flux causing a quenched-in density increase. The bonding hybridization is expected to adjust itself to the local density, becoming more sp^2 at low density and more sp^3 at high density. A high density requires an incident ion to penetrate the first atomic layer of the film and, temporarily, enter an interstitial position. Accumulation of these ions will increase the local density, provided there is little relaxation. The local bonding will reform around the atoms to become bulk bonding of the appropriate hybridization. If the ion energy is too low, the ion fails to penetrate, and it will just stick to the surface and form a sp^2 a-C.

At higher energies, the ion can penetrate further into the solid and increase the density in deeper layers. All penetrating ions will increase the density, in principle. However, the ion must also dissipate its kinetic energy. This energy dissipates quite rapidly in about 10^{-12} s in a 'thermal spike', but it is large and available to activate a relaxation of the excess density. The greater this excess energy, the greater the probability of relaxation. Hence, the optimum ion energy for maximum density is just above the penetration threshold.

Consider a flux F of C ions incident on a growing a-C film (figure 4). Let a fraction f of ions penetrate the surface layer and fraction $(1-f)$ remain on the surface to form sp^2 bonded a-C of density ρ_0. For simplicity, let the ions have a constant range R rather than a distribution. During a time Δt, the film will grow outwards by $\Delta x = F(1-f)\,\Delta t/\rho_0$, while a mass $Ff\Delta t$ of ions is deposited over the same distance Δx, at a depth R below the surface. This increases the local density there by

$$\Delta\rho = \rho_0 f/(1-f). \tag{3}$$

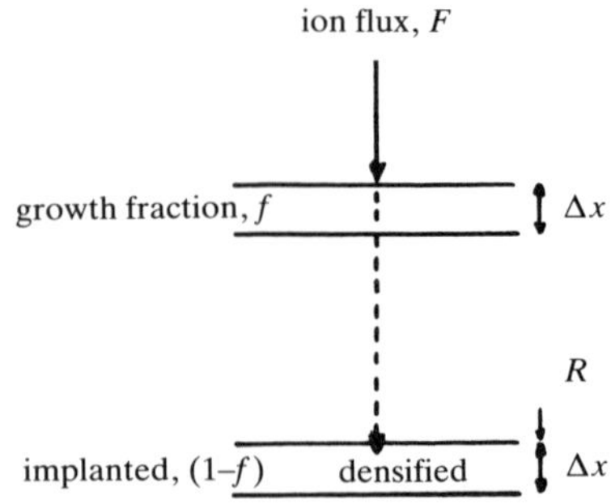

Figure 4. Proposed mechanism for promoting densification and sp^3 bonding by ion beam deposition in a-C.

Now suppose the film density at R can relax during the thermal spikes associated with each ion impact. Let each impact activate n_r atoms for relaxation. The fractional relaxation β is proportional to n_r and β itself, giving $\beta = 1-\beta n_r$ and $\beta = 1/(1+n_r)$. Thus, the density increase including relaxation is

$$\Delta\rho = \rho_0 \beta f/(1-\beta f). \tag{4}$$

This increase sweeps through the film at a depth R below the surface. The same increase occurs if the ion range is a distribution rather than a delta function, the increase is just an integral over the range. An upper layer of depth R remains unconverted with a density ρ_0.

Now consider the ion processes. The ion energies of interest (20–200 eV) are low. At this energy, the ion range is only a few monolayers and energy loss occurs mainly by nuclear stopping. Nuclear stopping consists of elastic collisions between atoms, in which an energy T is transferred to the target atom. The atom is displaced if T exceeds the displacement threshold E_d. E_d is the energy needed to create a permanent Frenkel (vacancy–interstitial) pair by displacement. This exceeds the isothermal Frenkel creation energy by an energy lost in migration and needed to separate the vacancy and interstitial sufficiently to prevent recombination. The Frenkel energies of graphite and diamond are about 15 eV and 21 eV respectively (Thrower & Mayer 1978; Bernholc *et al.* 1988), about half of E_d.

Ion penetration can occur in two ways. First, the ion can pass directly through the surface layer and enter an interstitial site (figure 5*a*). A threshold exists for this process because the collision cross-section increases rapidly at low ion energies. The cross-section diameter exceeds the interatomic separation for ion energies below about 20 eV. The ion range can be calculated by the TRIM code (Biersack & Haggmark 1980). However, this estimate may be questioned as many of the approximations used in TRIM break down at such low energies (Dodson 1990).

Ion penetration can also occur by displacement of an atom in the first layer (figure 5*b*). This requires the ion energy in the solid to exceed E_d. However, the surface is also an attractive potential barrier of height E_B for ions in the solid. E_B the surface binding energy equals the sublimation or cohesive energy, 7.4 eV for C. Thus the net penetration threshold is

$$E_p = E_d - E_B \tag{5}$$

or 28 eV, from $E_d = 35$ eV, $E_B = 7.4$ eV.

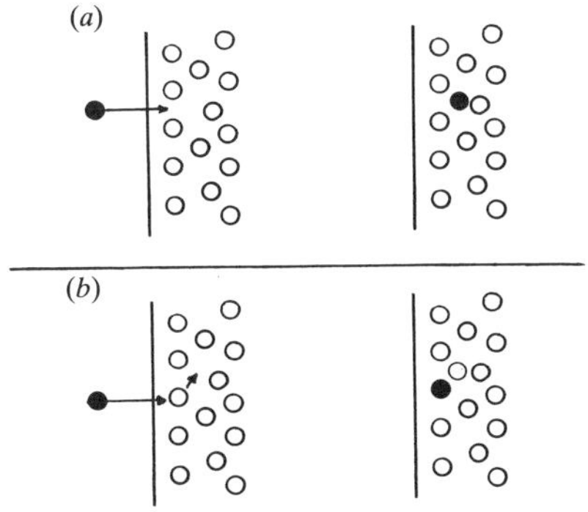

Figure 5. Impact of a low energy ion on a-C, showing (*a*) direct penetration and (*b*) penetration by displacement.

The relaxation of density can be described by the thermal spike model (Seitz & Koehler 1956). Consider a point source of energy Q dissipating by thermal conductivity. The temperature profile is given by

$$T(r,t) = Q/(8\pi^{\frac{3}{2}} c\rho[Dt]^{\frac{3}{2}}) \exp(-r^2/4Dt), \tag{6}$$

where $D = \kappa/c\rho$, ρ is the density, c is the specific heat and κ is the thermal conductivity. Suppose that annealing occurs by a thermal activated process with an attempt frequency

$$v = v_0 \exp(-E'/kT). \tag{7}$$

The total number of jumps within a spike (that is per incident ion) is given by

$$n_{\mathrm{r}} = \int_0^\infty \int_0^\infty v_0 \exp(-E'/kT(r,t))\, 4\pi r^2 \,\mathrm{d}r\, \mathrm{d}t. \tag{8}$$

Integration gives $n_{\mathrm{r}} = \{16\Gamma(2/3)/81\pi^{\frac{1}{3}}\} v_0 a^{\frac{2}{3}} (Q/E')$, where $a = (kQ)/(E'c\rho[4\pi D]^{\frac{3}{2}})$, or

$$n_{\mathrm{r}} \approx 0.016(v_0 r_s^2/D)(Q/E')^{\frac{5}{3}}, \tag{9}$$

where $c\rho \approx 3n_0 k$ and $1/n_0 = 4\pi r_s^3$. Most processes occur at early times in the spike when the temperature is very high. Thus, the phonon mean free path is of order of the interatomic spacing r_s, so $D \approx v_0 r_s^2$. This gives

$$n_r \approx 0.016(Q/E')^{\frac{5}{3}}. \tag{10}$$

The ion range at these energies is so low that each ion can be considered to produce a single thermal spike so that $Q \approx E_{\mathrm{i}}$. MSIB a-C has a relatively low thermal stability, transforming below $T_{\mathrm{t}} \approx 800$°C.This gives $E' = kT_{\mathrm{t}} \ln(v_{\mathrm{t}}) \approx 3.1$ eV, where $v \approx 10^{14}$ s^{-1} and $t \approx 1$ s, the experimental timescale. Note that E' is quite low compared with the C–C bond energy. The density increment can then be calculated using (4) and compared with experiment in figure 3. The agreement is reasonably good, considering the approximations used. The optimum ion energy is found to be $E_{\mathrm{i}} \approx 60$ eV, higher than found by McKenzie *et al.* (1991*a*).

The intrinsic stress of MSIB a-C arises from the high site energy of sp^3 a-C. Kelires (1991) found that the free energy of sp^3 sites in a-C was about 0.3 eV higher

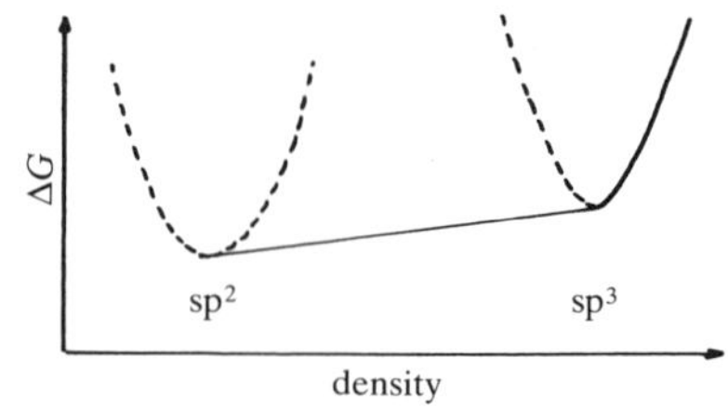

Figure 6. Schematic free energy against density for a-C.

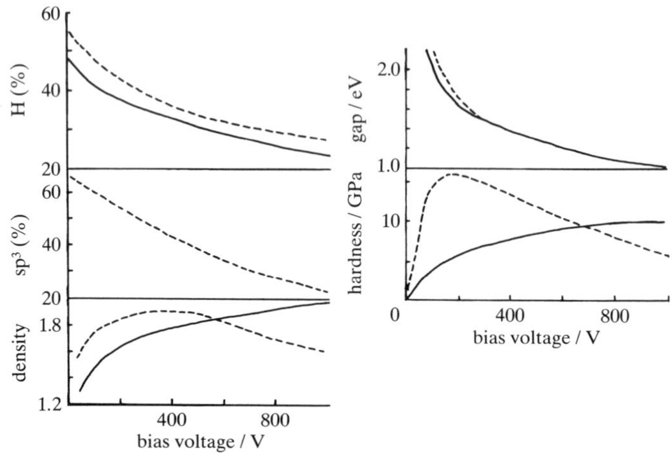

Figure 7. Variation of sp^3 content, H content, density, optical gap and micro-hardness of PD a-C:H deposited from methane (---) and benzene (——) (3 Pa pressure); data from Tamor *et al.* (1989, 1991), Koidl *et al.* (1990) and Jiang *et al.* (1989).

than sites in sp^2 a-C network or than diamond and graphite. Figure 6 shows schematically the free energy ΔG of a-C as a function of density. The dashed lines indicate the elastic response of a frozen network, while the full line indicates the ability of a continuously adjustable network to pass from one state to the other. The ion-bombardment model above suggests that the compressive stress P can be viewed as the work done by incident ions in compressing sp^2 sites into sp^3 sites in a disorder network. Thus

$$W = \Delta G = P\,\Delta V. \tag{11}$$

The stress is $P = 15$ GPa for $\Delta G = 0.3$ eV, $\rho_{sp^2} = 2.27$ g cm^{-3}, $\rho_{sp^3} = 3.51$ g cm^{-3}. Smaller density changes give proportionately lower observed stresses.

(*b*) PD a-C:H

The properties of plasma-deposited a-C:H depend primarily on the bias voltage and thus on the mean ion energy E_i. This dependence arises from the variation of the sp^3 and H content with bias (figure 7). The sp^3 and H content both fall with increasing bias (Koidl *et al.* 1990; Tamor *et al.* 1991). This leads to three regimes. At low bias, the high sp^3 and H content produces a polymeric solid, 'soft a-C:H', which is quite soft, light and has a wide band gap. At intermediate bias, the fraction of quaternary carbon (unhydrogenated sp^3 sites) reaches a maximum, giving films of 'hard a-C:H' of the highest density and hardness. At still higher bias, sp^2 bonding becomes dominant which leads to a fall off in density and hardness and a rapid closing of the band gap.

Koidl *et al.* (1990) found that the properties of soft a-C:H depended strongly on source gas, but those of hard a-C:H were not. Koidl attributed this to the complete decomposition of the source gas at high bias, which was incomplete at low bias. This conclusion was partly based on optical emission studies which found a strong signature of the CH radical in both benzene and methane plasmas (Wild & Koild 1987). However, it is now realized that the CH radical is unusually emissive, which can give misleading impressions of plasma dissociation. In fact, Robertson (1991*b*, 1992*a*) found that some properties like density depend strongly on source gas in both soft and hard regimes, while some properties like band gap are almost independent of gas in both regimes, as seen in figure 7.

An alternative model of plasma deposition is now proposed. We first note that while the degree of ionization in the plasma can be high, the degree of decomposition of the source gas is relatively low. Thus, the major ionic species detected by mass spectrometry in ion beams from CH_4, C_2H_4 and C_6H_6 plasmas are CH_3^+, $C_2H_2^+$ and $C_6H_6^+$ ions respectively (Weiler *et al.* 1992; Ehrhardt *et al.* 1992; Schaarschidt *et al.* 1991). Thus, the predominant ion contains the same number of carbon atoms as the source gas molecule. Ions are the major growth species in PD because they have a sticking coefficient of 1. Neutrals do make some contributions to growth, as is clear from comparing growth rates to ion fluxes (see, for example, Locher *et al.* 1991), but we neglect this for the moment. On impact, the molecule ion $C_mH_n^+$ will dissociate into m separate C^+ ions, the energy will be partitioned mainly between the daughter C^+ ions, giving each an energy E_i/m.

Ion bombardment is used to dehydrogenate a-C:H by the preferential sputtering of hydrogen, which is relatively weakly bound. Thus we expected the variation of hydrogen and sp^3 fraction on V_b to scale with source gas as $1/m$. This dependence is indeed found: the maximum density occurs at about $V_b = 300$ eV for a source gas of methane, 500 eV for acetylene and over 1000 eV for benzene (figure 7).

In contrast, the bias dependence of band gap is independent of source gas. In this case the band gap depends on the sp^2 site fraction and their clustering (Robertson & O'Reilly 1987). sp^2 sites are formed mainly by thermally activated elimination

$$=\mathrm{CH}{-}\mathrm{CH}{=} \rightarrow {=}\mathrm{C}{=}\mathrm{C}{=} + \mathrm{H_2}.$$

The sp^2 sites are likely to be formed by this reaction in a thermal spike. Now, although a molecular ion may dissociate on impact, partitioning its energy between m daughter ions, each ion will lose its energy within the same small region, within the same thermal spike. Thus, we expect properties depending on the sp^2 sites to be independent of m of the source gas, as is indeed seen in figure 7.

The author is very grateful to Dr A. Fuchs and Professor H. Ehrhardt for discussions and performing TRIM calculations.

References

Angus, J. C. & Hayman, C. C. 1988 *Science, Wash.* **241**, 913.

Aksenov, I. I., Vakula, S. I., Padalka, V. G., Strelnitski, R. E. & Khoroshikh, V. M. 1980 *Soviet Phys. tech. Phys.* **25**, 1164.

Bernholc, J., Antonelli, A., DelSole, T. M., BarYam, Y. & Pantelides, S. T. 1988 *Phys. Rev. Lett.* **61**, 2693.

Biersack, J. P. & Haggmark, L. G. 1980 *Nucl. Instr. Methods* **174**, 257.

Davanloo, F., Lee, T. J., Jander, D. R., Park, H. & Collins, C. B. 1992 *J. appl. Phys.* **71**, 1446.

Dodson, B. W. 1989 *Mater. Res. Soc. Symp. Proc.* **128**, 137.

Ehrhardt, H. *et al.* 1992 *Diamond Related Mater.* **1**, 316.

Fink, J. *et al.* 1983 *Solid State Commun.* **47**, 887.

Gaskell, P. H., Saeed, A., Chieux, P. C. & McKenzie, D. R. 1991 *Phys. Rev. Lett.* **67**, 1286.

He, H. & Thorpe, M. F. 1985 *Phys. Rev. Lett.* **54**, 2107.

Ishikawa, J., Takeiri, Y., Ogawa, K. & Takagi, T. 1987 *J. appl. Phys.* **61**, 2509.

Jiang, X., Reichelt, K. & Stritzker, B. 1989 *J. appl. Phys.* **66**, 5805.

Kaukonen, H. P. & Nieminen, R. M. 1992 *Phys. Rev. Lett.* **68**, 620.

Kelires, P. C. 1992 *Phys. Rev. Lett.* **68**, 1854.

Koidl, P., Wild, C., Dischler, B., Wagner, J. & Ramsteiner, M. 1990 *Mater. Sci. Forum* **52**, 41.

Koike, J., Parkin, D. M. & Mitchell, T. E. 1992 *Appl. Phys. Lett.* **60**, 1450.

Koskinen, J. 1988 *J. appl. Phys.* **63**, 2094.

Lifshitz, Y., Kasi, S. R., Rabalais, J. W. & Eckstein, W. 1990 *Phys. Rev.* B**41**, 10468.

Locher, R., Wild, C. & Koidl, P. 1991 *Surf. Coatings Technol.* **47**, 426.

McKenzie, D. R. *et al.* 1991*a* *Diamond Related mater.* **1**, 51.

McKenzie, D. R., Muller, D. & Pailthorpe, B. A. 1991*b* *Phys. Rev. Lett.* **67**, 773.

O'Keeffe, M., Adams, G. B. & Sankey, O. F. 1992 *Phys. Rev. Lett.* **68**, 2325.

Oelhafen, P., Ugolini, D., Schelz, S. & Eitle, J. 1991 In *Diamond and diamond-like carbon films* (ed. R. E. Clausing *et al.*), p. 377. Plenum.

Pan, H., Pruski, M., Gerstein, B. C., Li, F. & Lannin, J. S. 1991 *Phys. Rev.* B**44**, 6741.

Pailthorpe, B. A. 1991 *J. appl. Phys.* **70**, 543.

Robertson, J. 1986 *Adv. Phys.* **35**, 317.

Robertson, J. 1991*a* In *Diamond and diamond-like carbon films* (ed. R. E. Clausing *et al.*), p. 331. Plenum.

Robertson, J. 1991*b* *Prog. Solid State Chem.* **21**, 199.

Robertson, J. 1992*a* *Surf. Coatings Technol.* **50**, 185.

Robertson, J. 1992*b* *Phil. Mag.* B**66**, 199.

Robertson, J. 1992*c* *Phys. Rev. Lett.* **68**, 220.

Robertson, J. 1992*d* *Diamond Related Mater.* **1**, 397.

Robertson , J. & O'Reilly, E. P. 1987 *Phys. Rev.* B**35**, 2946.

Savvides, N. 1990 *Mater. Res. Forum* **52**, 407.

Schaarschidt, G. *et al.* 1991 *Mater. Sci. Engng* A**140**, 788.

Seitz, F. & Koehler, J. S. 1956 In *Solid state physics* (ed. F. Seitz & D. Turnbull), vol. 2, p. 305.

Spencer, E. G., Schidt, P. H., Joy, D. C. & Salasone, F. J. 1976 *Appl. Phys. Lett.* **29**, 118.

Steffen, H. J., Marton, D. & Rabalais, J. E. 1992 *Phys. Rev. Lett.* **68**, 1726.

Tamor, M. A., Haire, J. A., Wu, C. H. & Hass, K. C. 1989 *Appl. Phys. Lett.* **54**, 123.

Tamor, M. A., Vassell, W. C. & Carduner, K. R. 1991 *Appl. Phys. Lett.* **58**, 592.

Thrower, P. A. & Mayer, R. M. 1978 *Phys. Status Solidi* A**47**, 11.

Weiler, M., Kleber, R., Jung, K. & Ehrhardt, H. 1992 *Diamond Related Mater.* **1**, 121.

Wild, C. & Koidl, P. 1987 *Appl. Phys. Lett.* **51**, 1506.

Zou, J. W., Reichelt, K., Schmidt, K. & Dischler, B. 1989 *J. appl. Phys.* **65**, 3914.

Zou, J. W., Schmidt, K., Reichelt, K. & Dischler, B. 1990 *J. appl. Phys.* **67**, 487.

10

Applications of diamond-like carbon thin films

BY ALAN H. LETTINGTON

There is considerable interest in growing diamond layers and thin films for a wide range of applications. Some of these thin film requirements can be met already using existing diamond-like carbon coatings. This paper reviews the use of diamond-like carbon films in infra red optical, mechanical, electronic and biomedical applications.

---

1. Introduction

Carbon can exist in many forms, amorphous, glassy and crystalline. It is well known that at ordinary temperatures and pressures graphite is the thermodynamically stable crystalline form of carbon. Diamond can be formed at pressures of 30 kbar† and although it is metastable at atmospheric pressures it reverts to graphite only at elevated temperatures.

The diamond form has quite remarkable physical properties. It is exceedingly hard, has a high Young's modulus and tensile strength combined with high thermal conductivity and thermal shock properties (see table 1). It is chemically inert and in pure form has excellent infra red transmission. There is considerable interest in growing this material at low temperatures and pressures, either as a bulk material, or else as a protective coating on existing materials.

At present the optimum temperature for growing chemical vapour deposition (CVD) diamond is around 900 °C and although this may be lowered to around 600 °C it is still too high for many substrates. In those applications where thin protective layers are adequate then diamond-like carbon (DLC) may provide an immediate solution.

2. Preparation of DLC films

Diamond-like carbon does not have a unique composition but consists of a mixture of amorphous and crystalline phases. Its properties vary considerably with deposition conditions. There were a number of early reports of hard carbon films but most of the current interest was stimulated by the work of Aisenberg & Chabot (1971). They used a carbon ion beam source and accelerated carbon ions towards the substrate by means of a negative bias. They were able to produce films exhibiting many of the properties of diamond and called them diamond-like.

The preparation of similar films from a hydrocarbon gas in a DC plasma was reported by Whitmell & Williamson (1976) and in an RF plasma by Holland & Ohja

† 1 bar = 10^5 Pa.

Table 1. *Physical properties of diamond*

hardness	10 mg mm^{-2}
Young's modulus	945 GN m^{-2}
thermal conductivity	20 W cm^{-1} K^{-1}
refractive index	2.42
bandgap	5.45 eV

(1976). Since then DLC films have been produced by a variety of techniques including a dual-beam method (Weissmantel 1977), ion beam plating (Bewilogna *et al.* 1979), a simultaneous sputtering and RF plasma CVD process (Green & Lettington 1980; Zelez 1983) and by magnetron sputtering (Savvides & Window 1985).

The essential feature of all of these processes is that the films are grown under ion bombardment and Weissmantel *et al.* (1980) have proposed to call these films i-C for this reason. Many of the deposited layers contain significant amounts of hydrogen (see Angus *et al.* 1980, and enclosed references) and it has been proposed (Bubenzer *et al.* 1983) that these particular materials are called hydrogenated a-C or a-C:H, where a-C stands for amorphous carbon by analogy with amorphous silicon (a-Si). I propose to retain the original name (DLC) introduced by Aisenberg & Chabot (1971).

Diamond-like carbon films contain carbon atoms in a variety of different coordinations. There are the tetragonally coordinated sp^3 carbon atoms present in pure diamond as well as the trigonal sp^2 coordination as found in graphite and possibly some sp^1 coordinated atoms. DLC films may contain microcrystalline diamond and graphite as well as a disordered structure containing a mixture of configurations.

The physical properties of these films depend on the method of preparation. For a given method, the ratio of sp^3 to sp^2 coordinated carbon can be affected by the presence of hydrogen in the films. This ratio has been studied by NMR and XPS (Jansen *et al.* 1985; Grill *et al.* 1987, 1988) as well as IR spectroscopy of hydrogenated material (Dischler *et al.* 1983; Nadler *et al.* 1984). The results differ widely with method of deposition. In the case of the IR spectroscopy measurements, it is not possible to study sp^3 bonded material not containing hydrogen. Also it is difficult to identify diamond crystallites in DLC by Raman scattering but a number of workers have reported the presence of diamond from X-ray scattering measurements. Recently C. T. Pillinger, R. D. Ash, S. A. Russel and J. W. Arden (personal communication) treated a sample of DLC prepared by RF discharge with perchloric acid and obtained at least a 90% weight loss. The residue was studied by an SEM and found to contain a substantial number of individual well-formed octahedral crystals of diamond.

The metastable form of DLC may arise from the energy of the incident ions causing thermal and pressure spikes which are quenched in the depositing layer (see Grill (1988) for enclosed references).

3. Physical properties of DLC films

DLC films can be exceedingly hard. Values in excess of 3000 kg mm^{-2} have been reported (Weissmantel *et al.* 1979; Angus *et al.* 1986). The materials generally have a low coefficient of friction increasing with relative humidity (0.01–0.19) (Enke *et al.* 1980).

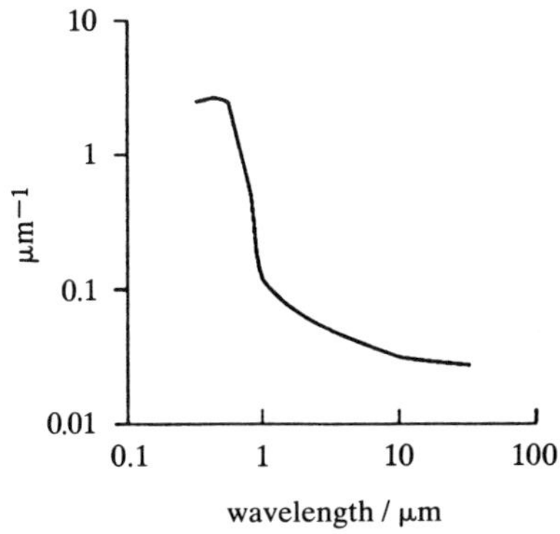

Figure 1. Absorption of DLC.

Layers prepared at high bias potentials have a high internal stress and thicknesses can be limited to 2–3 μm. Such films have been known to shatter germanium substrates and bow optical components.

DLC films are generally absorbing in the visible (see figure 1) but have good transmission in the infra-red region of the spectra and can be used for infra red antireflection coatings on germanium optics (Holland & Ohja 1979; Green & Lettington 1980; Bubenzer *et al.* 1983) and on silicon solar cells (Moravec & Lee 1982).

The coatings can have high electrical resistivity and are chemically durable and abrasion resistant.

4. Diamond-like carbon anti-reflection coatings on germanium

Initially 25 mm diameter test samples of germanium were coated and subjected to laboratory tests. The process was then scaled up to larger components which were subjected to various field trials. Finally the process was used to coat windows and lenses for project applications.

The laboratory testing was to RSRE Specification TS 1888 (Lettington 1985) and required the development of a special abrasion tester. The landbased trials involved a simulated windscreen wiper mounted on the outside of a vehicle undergoing extensive driving trials. The coating showed no sign of deterioration.

For sea trials two of our coated discs of germanium were mounted at sea level on a fort in the Solent along with an uncoated square of germanium. They remained at sea continuously for over four months. The coated samples were virtually unaffected while the uncoated sample was badly corroded and etched away.

The effect of rain impact damage on coated germanium samples was carried out at RAE Farnborough. They were exposed to their standard rain conditions for a few minutes at normal incidence to the rain (2.5 cm dm^{-3}) (Tattershall & Minter 1991) and at 87 m s^{-1} impact velocity. Pitting did occur but the extent was considerably reduced compared to uncoated germanium. The effect of rain damage can be reduced still further through inclining the window to the direction of flight. Our coated germanium samples have also been subjected to single liquid drop impact testing at Cambridge University (van der Zwagg & Field 1983).

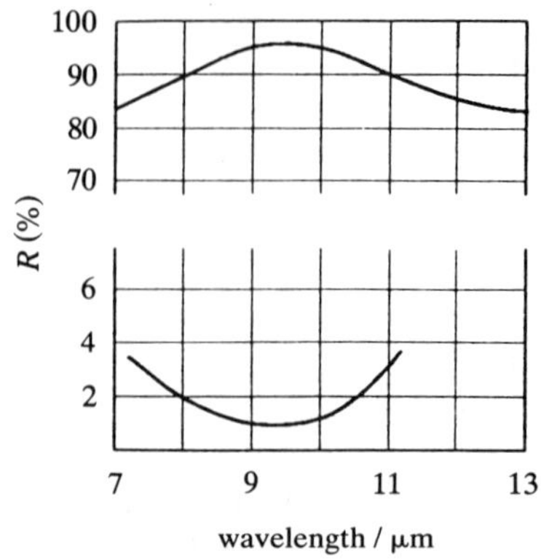

Figure 2. DLC coating on germanium.

A typical optical performance of the now commercially available DLC coating on germanium is illustrated in figure 2. It shows the transmittance of a 1 mm thick germanium substrate coated on the front surface with DLC to OCLI specification 6040011 and with a high efficiency AR coating 6040008, on the rear surface. There is a 5% loss in transmission at the peak wavelength due partly to an index mismatch and partly to absorption in the coating.

5. DLC coatings on zinc sulphide windows

Another of our coating requirements was to protect zinc sulphide infra red windows from rain impact damage. DLC coatings on their own were not adequate since the stress and absorption in the coating limited the thickness that could be used.

This problem was relieved to a certain extent through our development of GeC (Lettington *et al.* 1987). It is a tough durable coating material, with a refractive index a function of composition, that can be designed into multilayer structures. It bonds well to DLC, which is frequently used as the outermost layer. In this form it can act as a stress relieving coating for zinc sulphide windows. These coatings easily pass the RSRE sand/water wiper test of TS 1888.

6. Application of carbon coatings to front surface aluminium mirrors

In current thermal imaging systems, front surface mirrors produced by single point diamond machining of bulk aluminium are used for rotating polygons, flapping mirrors and for relay elements with optical power. The optical performance of these components tends to deteriorate with time and with exposure to the atmosphere. This process can be prevented through the use of a suitable optical coating. Unfortunately the reflectivity of these coated surfaces can be low when used at oblique incidence. This effect has been demonstrated (Cox *et al.* 1975; Lettington & Ball 1981) in aluminium mirrors protected with thin overcoatings of SiO_x and intended for use in the 8–12 μm spectral band. This effect occurs for only one direction of polarization, R_p, parallel to the plane of incidence. Similar effects are

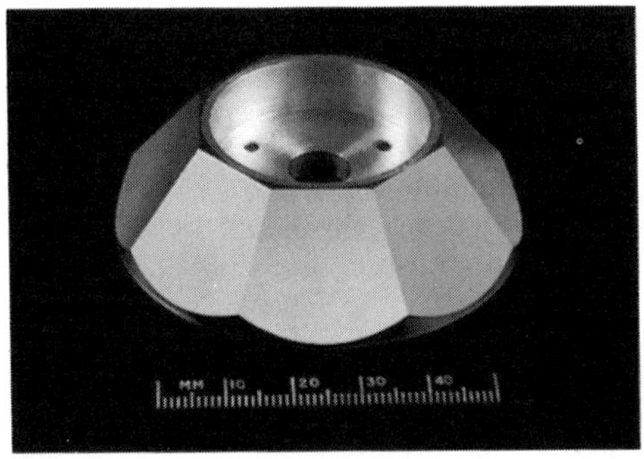

Figure 3. RSRE coated polygon for a coaxial scanner.

observed for many other protective coatings and other metallic reflectors (Pellicori 1978; Cox & Hass 1978) making these coatings unsuitable for use in 8–12 μm thermal imaging systems on 45° mirrors or scanning polygons.

The origin of this effect has been identified and it has been demonstrated that it does not occur in DLC protective coatings (Lettington & Ball 1981). The problem with coatings such as SiO_x is that they have strong optical absorption lines in the spectral region of interest. These coatings are sufficiently thin for this to produce negligible absorption. It does, however, affect the values of n and k in the absorbing region such that the Brewster angle at the air-coating interface occurs at very low angles of incidence. There is destructive interference between this reflection and that at the coating-metal boundary resulting in a loss of reflectivity. This effect does not occur with diamond-like carbon coatings on these mirror surfaces. An RSRE coated polygon for a coaxial scanner is illustrated in figure 3.

7. The use of DLC coatings for photothermal conversion of solar energy

The main aim of photothermal solar energy conversion is to collect solar radiation and to convert it into useful heat. There are two main types of converter: the flat plate collector, where an area of an absorbing material is placed so as to collect the solar radiation; and the focusing collector, where solar radiation is condensed on to a smaller absorbing area. Heat losses from the flat plate collector are high and in the U.K. the operating temperature is rarely higher than 70 °C. The focusing system has higher thermal efficiencies and operating temperatures of several hundred degrees celsius are possible. It is usual to remove the heat from both systems with circulating water.

For either collection process to be effective there must be maximum absorption α of the solar radiation and minimum heat losses. With regard to heat losses, convection and conduction are under the control of the system designer and must be minimized for maximum efficiency. It is the radiation loss ϵ from the absorber which is of consideration here. To maximize the ratio α/ϵ the surface must have high absorption (low reflectivity) from 0.3 μm to about 1.7 μm and low emission (high reflectivity) above 2 μm with a sharp transition between these two regions.

Several spectrally selective coatings have already been proposed and made using silicon or germanium layers deposited on to polished metal substrates (Hahn &

Seraphin 1978; Drummeter & Hass 1964; Seraphin 1979; Janai *et al.* 1979). The use of DLC for this purpose has been proposed by Ball & Lettington (1983). The absorption coefficient of DLC coatings was measured in the visible and infra red regions (see figure 1) and from this and the measured refractive index of about 2.2 values of α, ϵ and α/ϵ were calculated for C, Si and Ge single layer coatings of varying thickness deposited on to aluminium. A single layer of carbon has the highest efficiency. This value, however, it not high enough for most applications and we have sought ways to improve the α/ϵ ratio in multilayer designs.

8. Mechanical applications of DLC layers

In addition to the desirable infrared properties of DLC, discussed previously, the material is also hard and chemically durable making it useful for protecting metal objects from scratching and chemical attack. A variety of metal objects ranging from large sheets to nails, twist drills and machine tool inserts have been coated with DLC. Some of these have remained in the open exposed to the atmosphere for the past seven years without any sign of deterioration. A coated machine tool insert used for cutting aluminium at high speed lasted longer than uncoated inserts but the commercial viability of this coating process was not clearly established.

Other interesting features of DLC are the fact that it is hydrophobic and has a low coefficient of friction. We have coated a number of moving parts inside automobile engines and have successfully reduced wear rates. The frictional properties of DLC have been studied by Enke *et al.* (1980) using a ball-on-disc apparatus. They observed an increase in the coefficient of friction μ from 0.01 to 0.19 with increasing humidity. A sharp increase in μ occurs for relative humidities in excess of 1%. These results are contrary to those for graphite and diamond reported by Bowden & Young (1951) who observed a decrease in μ with increasing water vapour pressure. Similar but more detailed measurements have been reported recently by Kim *et al.* (1991) who included the effects of oxygen on the friction and wear of DLC films. They also considered the application of DLC films as protective overcoats on thin-film magnetic recording discs. Similar applications have been discussed by Tsui & Bogy (1987) and by Marchon *et al.* (1991) who correlated their observations with Raman and resistivity measurements. The application of DLC to magnetic and optical recording discs would appear to offer a promising market opportunity.

Another area in which we had success has been in coating thin optical fibres. A carbon coating inhibited the attack of the silica fibre by moisture so that brittle fracture was less likely to occur (see figure 4). These coatings were also able to restore the properties of aged fibres, probably as a result of the initial back sputtering stage removing some of the surface contamination.

9. Electronic device applications of DLC

DLC films have been studied as both active and passive elements in devices. Their use in an alternating current thin film electroluminescent device has been reported by Kim *et al.* (1990). The emission which occurs during breakdown of the DLC layer is very broad band, extending well into the UV and appears white. However, the brightness and efficiency of current devices are very low.

In another device application Kapoor *et al.* (1986) investigated the use of DLC films as the insulator in metal-insulator-semiconductor (MIS) devices. The results were not

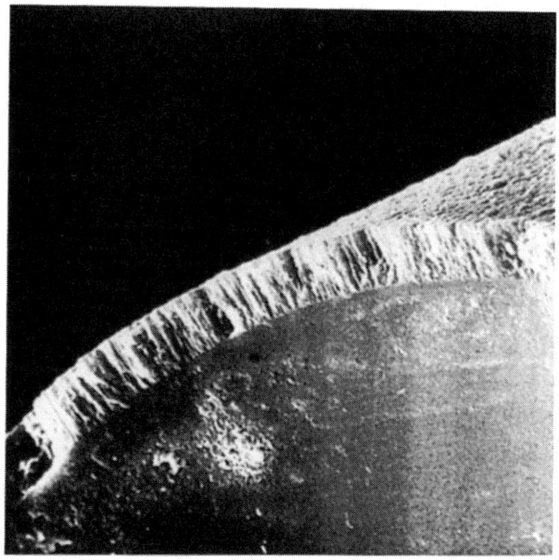

Figure 4. Cross-section of a coated optical fibre.

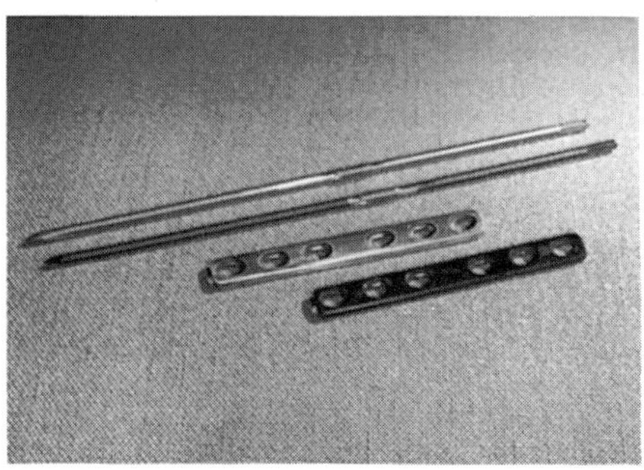

Figure 5. Carbon coated orthopaedic pins.

very encouraging owing to the low resistivity of the films and the large number of defects and traps present either in the layer or at its interface.

In a passive application Rothschild *et al.* (1986) have demonstrated the use of DLC film as a resist for high resolution photolithography of semiconductor surfaces. The process used an eximer laser to induce microchemical etching. A GaAs crystal was covered with a 0.2 μm layer of DLC and exposed to a 193 nm laser pulse. A sharp grating pattern was achieved after using a GaAs wet etchant.

The final device application to be considered uses DLC as an electrical insulator on copper heat sinks for logic and array chips (Marotta *et al.* 1991). The structure consists of an array of copper pistons that are spring loaded into an aluminium heat sink and press down onto the chips, which can dissipate up to 27 W each. The end of the copper piston in contact with the chips was coated first with nickel then amorphous hydrogenated silicon and finally with the layer of DLC. This coating prevented scratching of the copper that can occur normally on power on due to a mismatch of expansion coefficients. It also prevented electrical shorts without any deterioration in the thermal conductivity.

10. Medical applications of DLC layers

Carbon in various forms has been used for many biomedical applications during the past two decades (figure 5). Tissue can adhere well to carbon implants and sustain a durable interface. Also in the presence of blood a protein layer is formed which prevents the formation of blood clots at the carbon surface (Jenkins 1980).

Carbon fibre implants can promote the rapid ingrowth of tissue (Jenkins *et al.* 1977) and are used successfully for ligament repair. In bulk form porous charcoal permits a similar ingrowth but has low strength and may also present a site for infection. Carbon can be impregnated for example with resin to improve its properties but the ideal solution is probably to combine the strength of metals with the biocompatibility of carbon in the form of DLC coated metal implants.

Preliminary *in vitro* tests using mouse tissues have shown an encouraging degree of biocompatibility (Thomson *et al.* 1991) as have *in vivo* sheep tests conducted by Professor McGibben of Cardiff Royal Infirmary.

Far more laboratory testing needs to be done before these coatings can be accepted in human trials but if acceptable one could imagine coating many other implants such as the roots of false teeth.

References

Aisenberg, S. & Chabot, R. 1971 Ion-beam deposition of thin films of diamond-like carbon. *J. appl. Phys.* **42**, 2953.

Angus, J. C., Koidl, P. & Domitz, S. 1986 Carbon thin films. In *Plasma deposited thin films* (ed. J. Mort & F. Jansen). CRC Press.

Angus, J. C., Stultz, J. E., Schiller, P. J., Macdonald, J. R., Mirtich, M. J. & Domitz, S. 1984 Composition and properties of the so-called 'Diamond-like' amorphous carbon films. *Thin Solid Films* **118**, 311.

Ball, G. J. & Lettington, A. H. 1983 Diamond-like carbon coatings for the photothermal conversion of solar energy. *RSRE Memorandum no. 3617.*

Bewilogna, K., Dietrich, D., Paget, L., Schürer, X. & Weissmantel, C. 1979 Structure and properties of transparent and hard carbon films. *Surf. Sci.* **86**, 308.

Bowden, F. P. & Young, J. E. 1951 Friction of diamond, graphite, and carbon and the influence of surface films. *Proc. R. Soc. Lond.* A **208**, 444.

Bubenzer, A., Dischler, B., Brandt, G. & Koidl, P. 1983 RF-plasma deposited amorphous hydrogenated hard carbon thin films: Preparation, properties and applications. *J. appl. Phys.* **54**, 4590.

Cox, J. T., Hass, G. & Hunter, W. R. 1975 Infrared reflectance of silicon oxide and magnesium fluoride protected aluminium mirrors at various angles of incidence from 8 μm to 12 μm. *Appl. Opt.* **14**, 1247.

Cox, J. T. & Hass, G. 1978 Aluminium mirrors Al_2O_3 protected with high reflectance at normal but greatly decreased reflectance at higher angles of incidence in the 8–12 μm region. *Appl. Opt.* **17**, 333.

Dischler, B., Bubenzer, A. & Koidl, P. 1983 Bonding in hydrogenated hard carbon studied by optical spectroscopy. *Solid State Commun.* **48**, 105.

Drummeter, L. F. & Hass, G. 1964 *Phys. Thin Films* **2**, 305.

Enke, K., Dimigen, H. & Hubsch, H. 1980 Frictional properties of diamond-like carbon layers. *Appl. Phys. Lett.* **36**, 291.

Green, G. W. & Lettington, A. H. 1980 U.K. Patent Application GB 2069008.

Grill, A., Meyerson, B. S., Patel, V., Reimer, J. A. & Petich, M. A. 1987 Inhomogeneous carbon bonding in hydrogenated amorphous carbon films. *J. appl. Phys.* **61**, 2874.

Grill, A., Meyersen, B. S. & Patel, V. 1989 Interface modifications for improving the adhesion of a-C:H films to metals. *Proc. Soc. Photo-Opt. Instrum. Engng* **969**, 52.

Hahn, R. E. & Seraphin, B. O. 1978 Spectrally selective surfaces for photothermal energy conversion. *Phys. Thin Films* **10**, 1.

Holland, L. & Ohja, S. M. 1976 Deposition of hard and insulating carbonaceous films on an RF target in a butane plasma. *Thin Solid Films* **38**, L17.

Holland, L. & Ohja, S. M. 1979 The growth of carbon films with random atomic structure from ion impact damage in a hydrocarbon plasma. *Thin Solid Films* **58**, 107.

Janai, M. A., Allred, D. D., Booth, D. C. & Seraphin, B. O. 1979 Optical properties and structure of amorphous silicon films prepared by CVD. *Solar Energy Mater.* **1**, 11.

Jansen, F., Machonkin, M., Kaplan, S. & Hank, S. 1985 The effects of hydrogenation on the properties of ion beam sputter deposited amorphous carbon. *J. Vac. Sci. Technol.* A**3**, 605.

Jenkins, G. M. 1980 Biomedical applications of carbons and graphites. *Clin. Phys. Physiol. Meas.* **1**, 171–194.

Jenkins, G. M. & De Carvallo, F. X. 1977 Biomedical applications of carbon fibre reinforced carbon in implanted prostheses. *Carbon* **15**, 33.

Kapoor, V. J., Mirtich, M. J. & Banks, B. A. 1986 Diamond-like carbon films on semiconductors for insulated-gate technology. *J. Vac. Sci. Technol.* A**4**, 1013.

Kim, D. S., Fischer, T. E. & Gallois, B. 1991 The effects of oxygen and humidity on friction and wear of diamond-like carbon films. *Surf. Coatings Technol.* **49**, 537.

Kim, S. B. & Wager, J. F. 1990 Diamond-like carbon films for electroluminescent applications. *Surf. Coatings Technol.* **43**, 99.

Lettington, A. H. 1985 The development, application and testing of diamond-like coatings for infrared components. *Proc. Soc. Photo-Opt. Instrum. Engng* **590**, 100.

Lettington, A. H. & Ball, G. J. 1981 The protection of front surfaced aluminium mirrors with diamond-like carbon coatings for use in the infrared. *RSRE Memorandum no. 3295.*

Lettington, A. H., Lewis, J. C., Wort, J. C. H., Monachan, B. C. & Hope, A. J. N. 1987 Developments in GeC as a durable IR Coating Material. *E-MRS Meeting* **17**, 469.

Marchon, B., Heiman, N., Khan, M. R., Lautie, A., Ager, J. W. & Veirs, D. K. 1991 Raman and resistivity investigations of carbon overcoats of thin-film media: Correlations with tibological properties. *J. appl. Phys.* **69**, 5748.

Marotta, E., Bakhru, N., Grill, A., Patel, V. & Meyerson, B. 1991 Diamond-like carbon as an electrical insulator of copper devices for chip cooling. *Thin Solid Films* **206**, 188–191.

Moravec, T. J. & Lee, J. C. 1982 The development of diamond-like (i-carbon) thin films as antireflecting coatings for silicon solar cells. *J. Vac. Sci. Technol.* **20**, 338.

Nadler, M. P., Donovan, T. M. & Green, A. K. 1984 Structure of carbon films formed by the plasma decomposition of hydrocarbons. *Thin Solid Films* **116**, 241.

Pellicori, S. F. 1978 Infrared reflectance of a variety of mirrors at 48° incidence. *Appl. Opt.* **17**, 3335.

Rothschild, M., Arnone, C. & Ehrich, D. J. 1986 Eximer-laser etching of diamond and hard carbon films by direct writing and optical projection. *J. Vac. Sci. Technol.* B**4**, 310.

Savvides, N. & Window, B. 1985 Diamond-like amorphous carbon films prepared by magnetron sputtering of graphite. *J. Vac. Sci. Technol.* A**3**, 2386.

Seraphin, B. O. 1979 Chemical vapour deposition of spectrally selective surfaces.

Tattershall, P. & Minter, E. M. 1991 The rain field conditions of the DRA Aerospace Division whirling arm, rain erosion facility. *Technical Report 91063*, DRA Aerospace Division.

Thomson, L. A., Law, F. C., Rushton, N. & Franks, J. 1991 Biocompatibility of diamond-like carbon coating. *Biomater.* **12**, 37.

Tsai, H-C. & Bogy, D. B. 1987 Characterization of diamond-like carbon films and their application as overcoats on thin film media for magnetic recording. *J. Vac. Sci. Technol.* A**5**, 3287.

Van Der Zwagg, S. & Field, J. E. 1983 Indentation and liquid impact studies on coated germanium. *Phil. Mag.* A**48**, 767.

Weissmantel, C. 1977 In *Proc. 7th Int. Vacuum Congr. and 3rd Int. Conf. on Solid Surfaces*, p. 1533. Vienna: Berger.

Weissmantel, C. *et al.* 1980 Structure and properties of quasi-amorphous films prepared by ion beam techniques. *Thin Solid Films* **72**, 19.

Weissmantel, C., Schürer, C., Fröhlich, F., Grau, P. & Lehmann, H. 1979 Mechanical properties of hard carbon films. *Thin Solid Films* **61**, L.5.

Whitmell, D. S. & Williamson, R. 1976 The deposition of hard surface layers by hydrocarbon cracking in a glow discharge. *Thin Solid Films* **35**, 253.

Zelez, J. 1983 Low-stress diamond-like carbon films. *J. Vac. Sci. Technol.* A**1**, 305.

11

Diamond as a wear-resistant coating

BY B. LUX AND R. HAUBNER

Refractory chemical vapour deposition (CVD) coatings strongly improve the performances of hard metal tools. Low-pressure diamond synthesis permits both *in situ* CVD diamond coating and freestanding low-pressure diamond sheet fabrication. The performance of coated and freestanding bonded diamond layer tools approaches those of commercial polycrystalline diamond (PCD) products.

Low-pressure diamond grits and new composite powders could provide new and extraordinary grinding powders of raw materials for novel types of PCD composites.

Commercial production is now mainly a scaling-up problem. Reliable and low-cost fabrication are among the most important requirements. As these applications tolerate the use of polycrystalline layers, they are likely to be among those industrial low-pressure diamond products soon having a significant market share.

1. Importance of the new diamond technology

(*a*) *Introduction*

Thin-film and coating technologies now used to produce diamond can lead to important improvements in industrial applications in many fields (Lux & Haubner 1992). If integrated into the function of a system, the lifetime and technical performance of the resulting components, such as tools, etc., can be significantly increased. For many applications the benefits obtained through thin films far exceed their manufacturing costs. Even if these costs are high, the resulting increase in performance of the components for their applications is justified by highly competitive improvements. Also novel diamond products, impossible to fabricate till now, can be expected from this new technology in the near future (Lux & Haubner 1989).

A few selected examples will illustrate the principles and possibilities of this new technology.

(*b*) *Industrial applications of wear-resistant surface layers*

1. Improvement of materials by hard chemical vapour deposition (CVD) and polycrystalline diamond (PCD) coatings are widely used today. Single-phase layers (such as Al_2O_3, TiC, TiN, Ti(C, N), HfN, etc. (Lux & Schachner 1978; Lux *et al.* 1971, 1989; Holleck 1986)) and multilayers (consisting of several single layers deposited sequentially, these are known as composite coatings) improved considerably many industrial cutting tool applications. Hard metals and ceramic materials are frequently used as substrates (Lux *et al.* 1987).

2. Superhard high-pressure products (i.e. PCD and polycrystalline cubic boron nitride (PCBN)) currently have a well-established market (Heath 1989). Their properties and performances under severe working conditions are generally

considered to be outstanding and highly competitive. They are the ultimate standard to be reached or, if possible, even to be exceeded by the new low-pressure diamond products now being developed.

Both PCD and PCBN are sintered products with appropriate amounts of a binder, and compacted at high temperatures and ultrahigh pressures. Frequently the PCD or PCBN layers are bonded directly to a hard metal substrate.

Applications of PCD tools (Heath 1989; Schweitzer 1989) are primarily in cutting, chipless forming, and wear part applications of abrasive workpiece materials such as: wood-based products; non-ferrous alloys (Al–Si, Cu, etc.); green ceramic and hard metals; abrasive reinforced plastics, man-made resin-bonded wood particle boards, plastic-laminated chip boards; stone and rock drills; wire-drawing dies; and other general applications of wear-resistant surface layers. Cutting speeds and conditions can be very high and severe (Ebberink 1991).

PCBN tools (Heath 1989) are preferably used for machining (turning, milling, boring, honing, etc.) and chipless forming of ferrous metals, such as nickel based superalloys (Inconel, Nimonic), gray and hard cast irons, hardened ferrous steels (greater than 45 R_c), special hard facing materials, etc.

Serious restrictions for certain wear applications exist.

(i) Ultrahigh-pressure technology leads to relatively expensive products.

(ii) Fabrication of three-dimensional (3D) shapes is limited and generally linked with time-consuming, complex, high-cost operations.

(iii) Because of the high-pressure technology all products have definite upper sizes and shape limits, which cannot be exceeded without prohibitive cost increases.

(*c*) *Conclusions*

Compared with the high-pressure techniques, the low-pressure approach offers more flexibility while remaining simpler, less expensive and appropriate for novel lower-priced mass production with good performances. New applications and broader marketing possibilities should result from this. Therefore the lower costs linked with equal, or even better, performances of the low-pressure technology as compared with the ultrahigh-pressure technology will be key factors.

2. State of the art

(*a*) *Low-pressure diamond layers as a challenge*

Property requirements of low-pressure diamond for wear applications

Superhard, wear-resistant protective coatings for tools and antiwear parts as well as decorative, scratch-resistant or corrosion-resistant surfaces need and utilize mainly the following outstanding properties of diamond: superhardness (high Young's modulus, rigidity, etc.); high degree of chemical inertness (antisticking surface, etc.); and high thermal conductivity.

These properties are relatively insensitive to lattice defects. Most applications tolerate polycrystallinity and even a certain amount of sp^2 bonding (Sato 1990) without major detrimental effects on the antiwear performances of the diamond. This greatly simplifies the development of industrial wear applications as compared with other uses (Badzian *et al.* 1987).

High quality low-pressure PCDs are facetted. Pure and binderless their crystal facets, orientation, internal defect structures, sp^2/sp^3 bond ratios and grain sizes can be controlled within certain limits by the synthesis parameters. Furthermore these

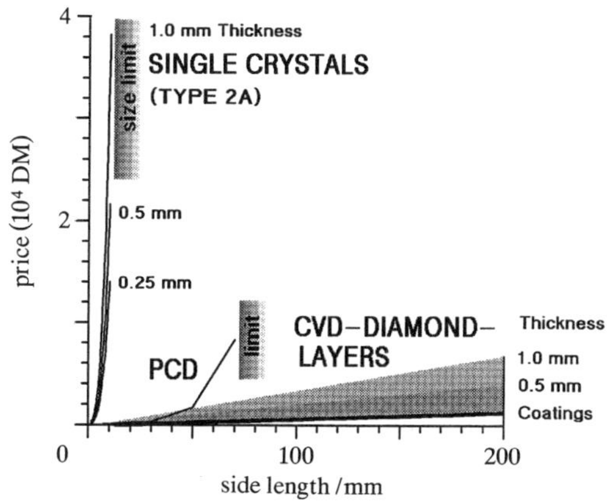

Figure 1. Comparison of the diamond price per surface area for different diamond products.

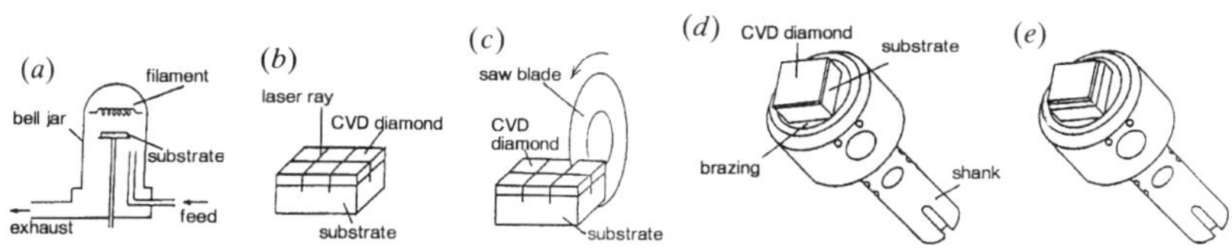

Figure 2. Production process of an *in situ* CVD diamond coated tool. An example is TAB bonding tool for micro electronics (Yashiki *et al.* 1991). (*a*) Synthesis of CVD diamond, (*b*) laser cutting, (*c*) saw blade cutting, (*d*) brazing, (*e*) CVD diamond TAB tools.

diamonds are available not only as bulky crystals but also in large two-dimensional shapes of defined thickness. Figure 1 shows a rough price comparison for different diamond qualities.

If used as an antiwear surface layer the low-pressure diamond permits the full utilization of diamond's extreme hardness, high modulus, thermal conductivity and other properties. The 'multilayer approach', today widely applied for conventional CVD cutting tools (Lux *et al.* 1987; Bichler *et al.* 1988; Schachner *et al.* 1984), should permit the development of tailor-made surfaces best suited for specific applications. It can also be applied to freestanding sheet products (Tanabe & Fujimori 1990).

Two different techniques are applied today for producing antiwear surface layers: *in situ* CVD coating (figure 2); and bonding of freestanding diamond sheets (figure 3). A technical comparison of these two products (figure 4) and the PCD products is given below.

In situ coating

Pros

(i) simple;
(ii) one-step operation for many substrates;
(iii) complex 3D shapes, due to high throwing power (tubings, dies, nozzles);
(iv) multilayers.

Cons

(i) chemical reactions of substrates with $C/H^0/H_2$;

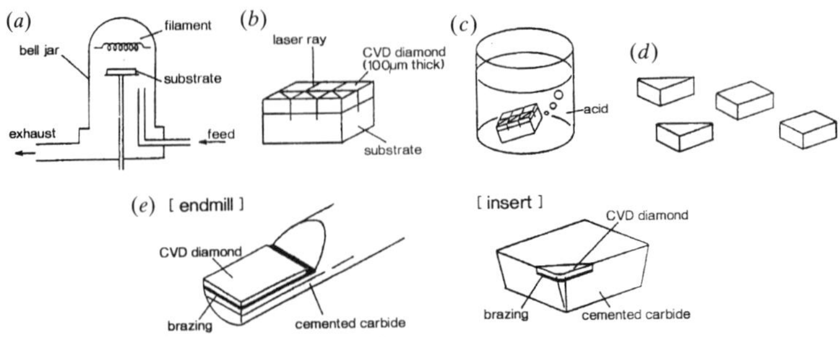

Figure 3. Production process of freestanding CVD diamond sheets brazed to cutting tools (Yashiki et al. 1991). (a) Synthesis of CVD diamond, (b) laser cutting, (c) dissolution of substrate, (d) freestanding CVD diamond pieces, (e) CVD diamond cutting tools.

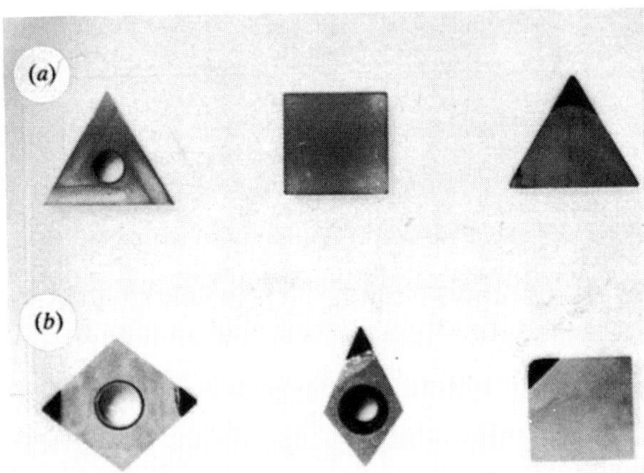

Figure 4. Examples of cutting tool inserts for metal cutting (Asahi Diamond Industrial). (a) *In situ* CVD coated, (b) tools with bonded freestanding CVD diamond sheets.

(ii) difficult adhesion with high expansion substrates;
(iii) rough as-grown surfaces;
(iv) substrate cooling for high growth rates.

Freestanding layer bonding

Pros (in comparison with *in situ*)
(i) separate processes for layer formation and bonding;
(ii) nucleation and growth on ideal substrates;
(iii) less interfacial stresses;
(iv) continuous production lines.

Pros (in comparison with PCD)
(i) binder-free;
(ii) large size sheets are no problem;
(iii) simple, inexpensive equipment;

Table 1. *Results of tool life tests and bonding cost analysis*

tool materials	CVD diamond	sumicrystal	PCD1	PCD2
tool life ($\times 10^3$ shots)	600	2000	200	80
maximum regrinds	—	3	3	3
total tool life ($\times 10^3$ shots)	600	8000	800	320
tool price (CVD unit base)	1	2.5	1	1
regrinding price (PCD unit base)	—	2.5	1	1
bonding cost (CVD unit base)	1	1	3.2	2.9

(iv) easy control of grain sizes, structure, etc.;
(v) multilayer sheets, allowing grades.

Cons (in comparison with *in situ*)
(i) only flat or simple 3D shapes;
(ii) many production steps (dummy surface preparation, diamond sheet fabrication, cutting, brazing to the tool, final grinding and polishing operations)

Cons (in comparison with PCD)
(i) grade nuances (due to missing binder)

(*b*) *Potential products and industrial applications*

In situ CVD diamond

TAB (tape automated bonding) stamps. These are used in microelectronics to connect chips to wires and circuits (Yashiki *et al.* 1991; Sumitomo Electric Industries 1990).

Recently, there has been a large demand to increase the through-put for flat packaging (identity cards, calculators, liquid-crystal displays, etc.) and to decrease production costs. The TAB process is now accepted by many electronic goods suppliers. During bonding the tools are heated to 500–600 °C and pressed down by a load.

CVD–TAB tools are special ceramics coated with a 50 μm thick hot-filament CVD diamond layer (figure 3). The diamond surface is polished until its surface roughness is less than 0.1 μm.

Both flatness and wear resistance tests of TAB tools are very important (Yashiki *et al.* 1991).

Technical results

Cleaning damage to CVD diamond is less than that of PCD, and CVD diamond showed a resistance as single crystalline diamond. As well as the economic advantage, low-pressure diamond further offers the possibility of producing much larger tool sizes than single-crystalline high-pressure diamond (Yashiki *et al.* 1991).

Cost assessment and lifetime

1. Cost per bond (including regrindings) for both CVD diamond tools and single-crystalline tools are considerably lower than those of PCD (table 1).

2. Single crystalline tools have the longest lifetime (eight million individual bonds) but high initial costs. They are most suitable for mass production with only slight variation in design.

3. CVD diamond and PCD tools show almost the same lifetime (several hundred thousand shots). They are best suited for medium-variety and medium-quantity or large-variety and small-quantity productions.

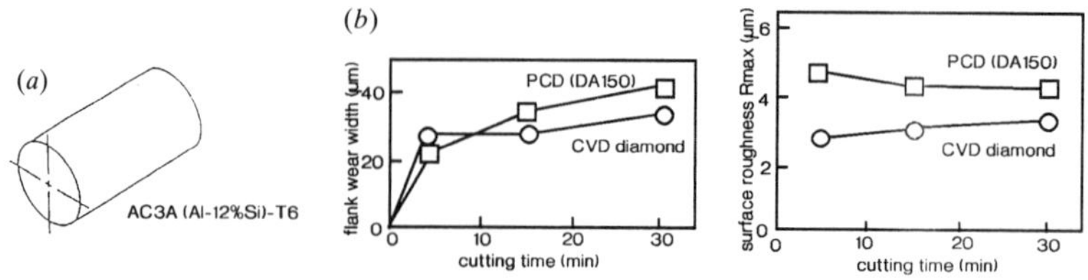

Figure 5. Comparison of flank wear width and surface roughness during dry continuous-cut turning. (*a*) Work piece, (*b*) results (Yashiki *et al.* 1991).

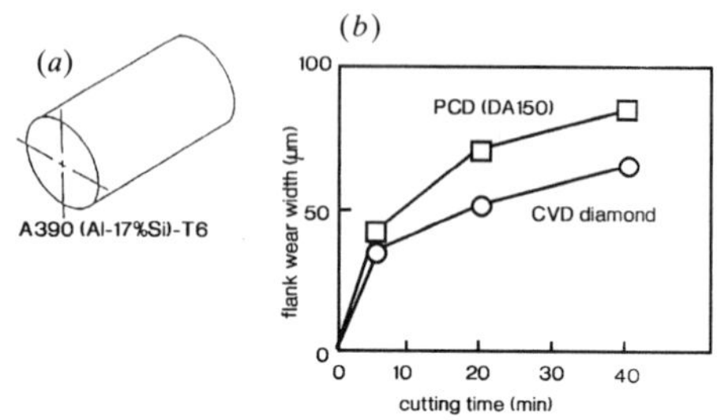

Figure 6. Comparison of the performance of PCD and CVD diamond layers during wet continuous-cut turning. (*a*) Workpiece, (*b*) results (Yashiki *et al.* 1991).

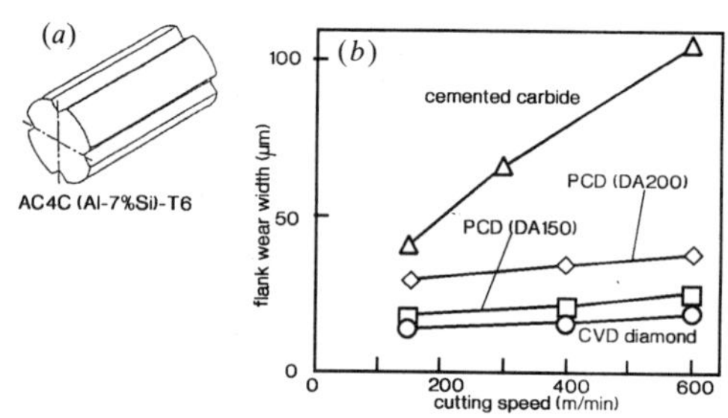

Figure 7. Comparison of the performance of hard metal and different diamond qualities during interrupted-cut turning. (*a*) Workpiece, (*b*) results (Yashiki *et al.* 1991).

Metal cutting tools

Continuous (turning) and interrupted cutting (milling) of Si rich hypereutectic Al–Si alloys under wet and dry conditions were examined recently by a number of research groups (Lux & Haubner 1989; Oakes *et al.* 1990; Yashiki *et al.* 1991).

Cutting performances of CVD diamond coated tools are excellent (figures 5–7) in comparison with hard metals; PCD can be reached in many cases (Yashiki *et al.* 1991; Sumitomo Electric Industries 1990).

Face milling and other factory machining tests. These showed that CVD diamond coated inserts can perform much better than hard metals (Yashiki *et al.* 1991). Also face milling of ceramic workpieces, copper alloys, graphite boards and wear-resistant grinding wheels was excellent (Kikuchi 1989).

Microdrills and twist-drills. These are made of hard metals (figure 8) are of great potential interest but more difficult to achieve.

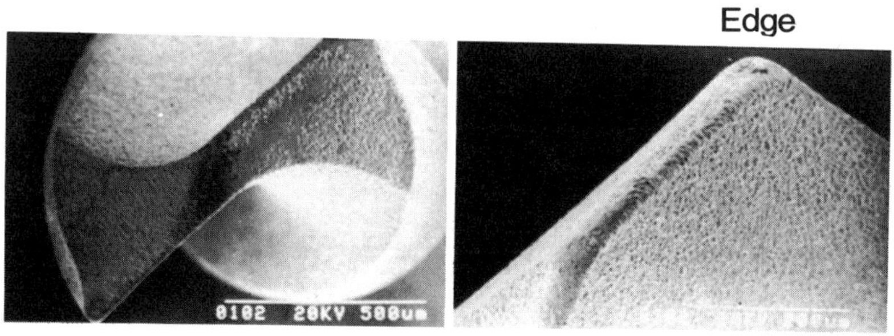

Figure 8. Microdrill *in situ* coated with CVD diamond (Asahi Diamond Industrial).

Industrial endmills and metal-cutting tool inserts. Those using bonded free-standing CVD diamond (figure 3) have been commercially available for a few years. They are simple to produce and have no cobalt or other binders in the diamond sheet, which is an advantage over PCD. During machining of copper-coated printed circuit boards (GFRP) they revealed clear superiority over hard metals. Tool life can be improved more than 10 times. Figure 9 shows the low flank wear observed with a diamond surface as compared with that for conventional hard metal endmills.

(c) *Conclusions*

Both techniques, *in situ* and bonded free standing layers, have their technological advantages and disadvantages which influence their price and their performance (see above). (For example, *in situ* diamond-coated cutting tool inserts, as well as free-standing endmill tools, are between those of PCD and hard metal tools and sometimes even superior to PCD.) It must be stressed that such test results are only excerpts from current research projects. Until now most results were obtained with specimens from experimental laboratory units (fewer than 20 specimens) or small prototype reactors (up to a few 100 specimens). Further important improvements in performance can be expected through an optimization of adhesion by a controlled substrate surface chemistry (surface preparations, minimized internal and interfacial stresses), and optimized coatings (structure, thickness, grain size, surface roughness, etc).

3. Technological gaps and currently unsolved problems

(a) *Problems related to fabrication and adhesion*

Influences of substrate chemistry, nucleation and growth.

Diamond nucleation on highly perfect substrate surfaces, for example silicon wafers, is sluggish and starts only after a long initial time period. Nucleation is also difficult if the surface carbon concentration is lowered rapidly by diffusion into the substrate (Lux & Haubner 1991).

Chemical reactions occurring between substrate and reaction gas (carbon, atomic and molecular hydrogen) can have an important influence on nucleation and adhesion (Lindlbauer 1989).

Nucleation enhancements, such as scratching the substrate surface with diamond powders, are generally effective in enhancing the diamond nucleation density (Lux & Haubner 1991). Without such enhancement it can be a major problem to obtain very thin coatings. Thick layers having facetted diamonds can lead to important

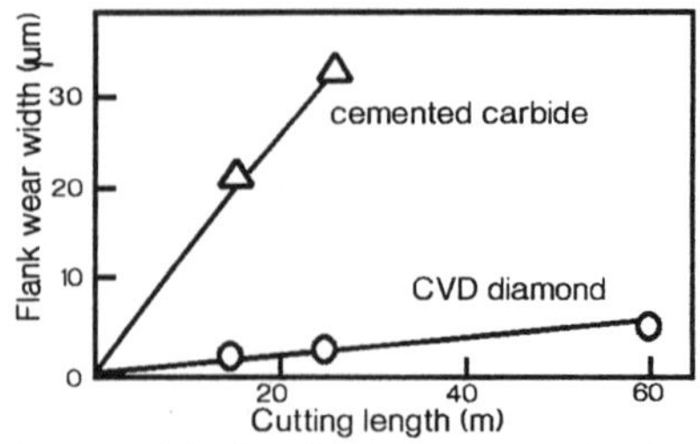

Figure 9. Wear resistance of double-flute CVD diamond endmill (Yashiki *et al.* 1991).

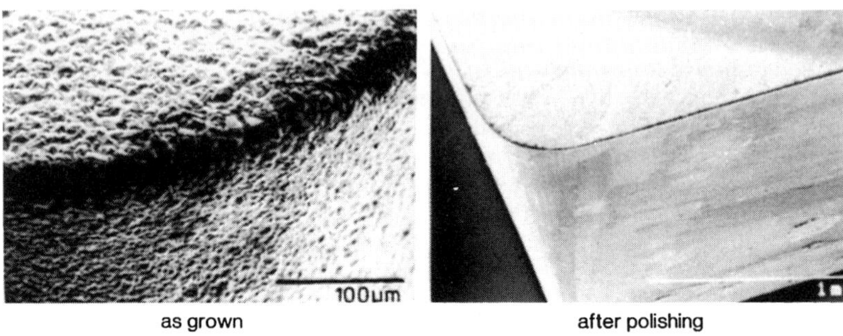

Figure 10. Roughness of CVD diamond. Coated cutting edge before and after polishing (Asahi Diamond Industrial).

surface roughnesses, undesirable for many applications. Costly grinding and polishing operations can become necessary (figure 10).

Fundamental problems of adhesion

Bad adhesion is frequently strongly associated with interfacial stress, which depends primarily on the different expansion coefficients of layer and substrate. Also excessive internal layer stress is important (Reineck *et al.* 1990).

The simple linkage of the substrate surface temperatures during deposition and the expansion coefficients of different substrates are shown in figure 11. It was shown recently that lowering the surface temperature during diamond deposition on hard metal tools can improve adhesion of the coating (Oakes *et al.* 1990). A strong chemical bonding between diamond and the substrate material is necessary for achieving the bondstrength needed for CVD coatings of cutting tools. Although the reproducibility still has to be improved, some diamond-coated substrates are already able to achieve the required excellency in adhesion for acceptable performances (see §2*b*).

However, chemical interactions leading to formation of solid intermediate layers or gaseous species can also be detrimental (Reineck *et al* 1990; Lux & Haubner 1991; Lindlbauer 1989).

(*b*) *Synthesis methods and industrial fabrication units*

Though during the past few years a large number of different new methods for low-pressure diamond deposition have been developed (Lux & Haubner 1992), mass production of *in situ* and freestanding layers for tools and wear applications still

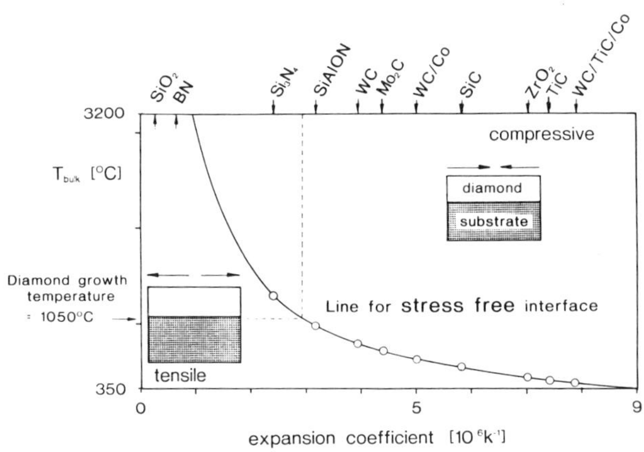

Figure 11. Influence of temperature and expansion coefficients on interfacial stress at a constant substrate surface temperature of 1050 °C (Lindlbauer 1992).

requires scaling up of the reactors. The most important technical factors for the choice of the synthesis method for industrial production are the reliable control of the deposition process (so as to obtain a constant and reproducible high-quality diamond) as well as the economy of the process. Size and shape of the products will largely determine the production units.

The following discussion considers mainly the problems of small mass products such as indexible inserts for cutting tools. In principle either a method with extremely high local growth rates (i.e. small deposition area), or a method that has a sufficiently large deposition surface/volume, with only a moderate local growth rate, can be chosen.

For *in situ* coating, in batch-type production reactors with low growth rate diamond methods, the deposition area must be large enough to ensure a reasonably high throughput. Only if the diamond growth rates are very high can shorter deposition times become possible, so that continuous production units might be feasible.

However, for reasons of control and reproducibility, lower growth rate methods seem preferable, as they allow an easier control of the diamond quality. Currently, hot-filament and microwave gas activation methods fulfil these requirements best and are now generally considered to be the preferred choice for scaling up units for the above products.

High growth rates are obtained with jet methods, such as plasma arc jets or the surprisingly simple oxygen/acetylene flame, both of which can work at atmospheric pressure. Deposits of high-quality diamonds were obtained with both methods.

The interesting synthesis possibility of halogenides was recently demonstrated (Patterson *et al.* 1991; Rudder *et al.* 1991) but is not yet well enough described for conclusions to be drawn. These methods could have a greater potential for inexpensive diamond grit.

Freestanding layers could be produced by both batch-type and continuous units. Most of the above considerations for *in situ* diamond production are also valid for freestanding layers, with the extremely important exception that for freestanding layers the optimal substrate for nucleation and growth can be freely chosen. 'Ideal

Table 2. *Comparison of the specific energy* (kWh g^{-1} *deposited*) *for a high- and a low-growth rate method of diamond deposition*

	microwave	arc jet
methane flow/($cm^3\ s^{-1}$)	1	80
power/kW	1	9.5
deposition rate/(μm h^{-1})	1	500
deposition area/cm^2	20	0.1
carbon supplied/(g h^{-1})	0.032	2.6
carbon deposited/(g h^{-1})	0.0071	0.0176
carbon conversion efficiency (%)	22	0.68
specific yield/(kWh g^{-1})	140	540

substrates' can be used because interface bonding requirements between dummy and diamond are much less critical than for the *in situ* coating and thus much easier to meet. Freestanding sheet materials are simply cut to the required shape and size needed for the final superhard surface before bonding.

Economical, continuous high-speed production of diamond-coated dummies can be envisaged, with diamond synthesis methods having high growth rates if several deposition sources are applied consecutively.

For thick sheets, highly polished dummy surfaces can produce a smooth replica diamond surface which can be used directly if the interface is turned inside out and used as the working surface. In connection with precision bonding it could minimize or possibly avoid the costs of the final grinding of the tool.

(*c*) *Economic considerations*

Whichever method is used it must fulfil not only the technical requirements noted above but also the economic ones. Despite the large number of methods available, the choices for industrial production methods are limited if economic and technical requirements are taken into account. Critical evaluations of the raw materials and the energy yielded, as well as the costs of the investments and labour are necessary. A comparison of two concepts for applying low-pressure diamond is given in table 2. An exact assessment demands a thorough knowledge of the processes and such information is still scarce. Recently L. K. Bigelow (personal communication) pointed out the importance of the specific energy per carat deposited. Comparing the two representatives of a high-growth and a low-growth rate method, he came to the conclusion given in table 2. Such important methods as hot-filament, DC plasma and flames are unfortunately not included in their considerations. It is clearly too early to give a full economic evaluation here.

4. Market potential and future prospects

(*a*) *Market potential*

Figure 12 shows the potential market for CVD diamond products for the year 2000. According to this estimate, from a total market of \$6500 M, thin-film products (*in situ* CVD coatings) will account for \$4500 M and the thick-film diamond products (freestanding films and sheet materials) are expected to amount to \$2000 M. Applications for wear parts, material-removal products and abrasives will be the most important markets sectors for CVD diamond in the near future.

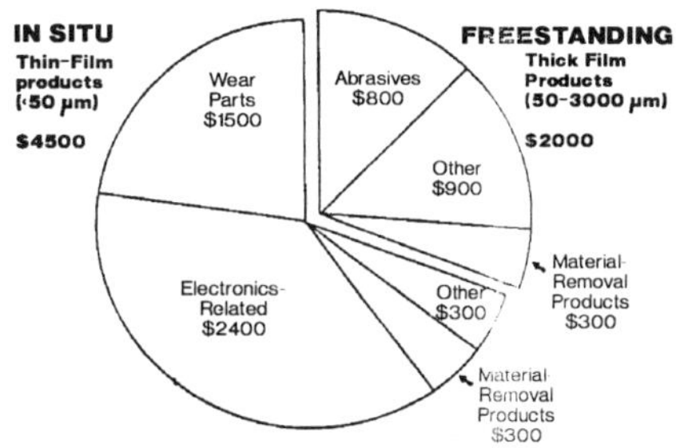

Figure 12. The potential market for CVD diamond products for the year 2000 (in million US dollars) (Genasystem 1990).

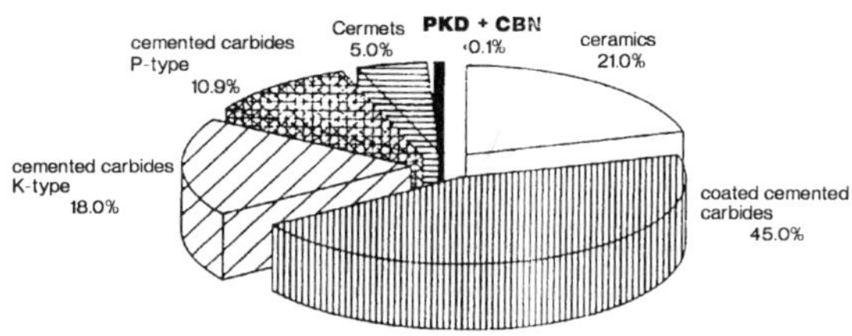

Figure 13. Typical distribution of cutting tool materials used in a modern automotive industry (Johannsen 1989).

The modern automotive industry is a good example of industrial demands (Johannsen 1989). Si-rich hypereutectic Al–Si alloys, used more and more for engine parts, have to be turned, milled, honed, etc. These specific applications are taken as an example and commented briefly in the following. Figure 13 shows that currently normal PCD and PCBN materials represent together just a small fraction (less than 0.1 %) compared with the other tool materials. The high percentage of CVD or PCD coated hard metal tools (multilayers of TiC, TiN, Al_2O_3, etc.) compared with the uncoated tools demonstrates the enormous impact the *in situ* CVD coating has had on tooling applications since it was first applied some 20 years ago (Lux & Schachner 1978).

Although the performance standard of PCD products is undoubtedly the goal to be reached, the object is not merely substitution of these products. To reach the full potential of CVD diamonds, much broader use is needed.

The future market share of low-pressure diamond cutting inserts and other tools will thus be based mainly on the technical advantages of the CVD technology, such as allowing the realization of more complicated tool shapes (J. Motzet, personal communication). In combination with the well-known benefits of PCD products there is a wide field of new market opportunities.

In the end, the decisive reason for a mass producer to choose low-pressure diamond tools will be the lower global costs per piece produced.

In this case the tool costs themselves are not the most important. A tool with the superiority of pure diamond – having a much greater lifetime and giving excellently machined surfaces with exact tolerances – means more reliable products with less efforts in man power.

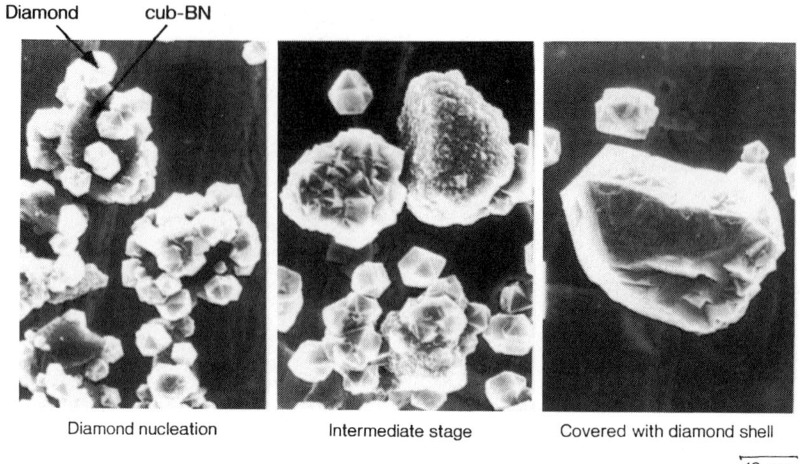

Figure 14. Different stages during production of a diamond composite powder (cub-BN core covered with a diamond shell) (Haubner 1990; Lux 1986).

Moreover, significantly higher cutting speeds and the higher chip volume produced per tool save production time per piece manufactured. They permit a more efficient use of highly complex, automated and expensive machines by reducing their non-productive time periods (less downtime for replacements of tools, adjustments, maintenance, etc.). This also lowers the investment costs for equipment and capital costs for building. Thus an enormous amount of money is free to pay for excellent tools that can substantially reduce production costs (J. Motzet, personal communication).

(*b*) *New superhard composite powders and compacted products*

Abrasive natural or high-pressure diamond powders have always held an appreciable share of the world's diamond market. Low-pressure diamond could become a competitor in future. Low cost production chemical processes, such as the recently discovered halide synthesis methods (Patterson *et al.* 1991; Rudder *et al.* 1991), might be of great interest.

Recently the feasibility of diamond 'composite powders', which consist of hard particles enveloped by a low-pressure diamond coating was demonstrated (figure 14). Such powders permit combinations of materials previously impossible to realize. Besides their use as abrasives, composite powders mixed with binders, compacted and sintered at ultrahigh pressures, or even conventionally compacted at low pressures by a classical sintering operation, could result in exciting, novel PCD-type products. Since abrasive products are relatively inexpensive, a low-pressure powder fabrication in large reactors (Litos 1992) with excellent yields is required to bring the production costs down to competitive levels. However, materials with outstanding properties and performances usable for unique, sophisticated applications are not subject to stringent price limitations (Lux 1986).

(c) The superhard BN phases

Many analogies exist for the boron nitride materials. The superhard BN phases are extremely interesting for wear applications. They offer an alternative set of diamond's chemical properties in such applications. As discussed for diamond, utilizing mainly the superhardness means that polycrystallinity and lattice imperfections are acceptable. Therefore, once appropriate low-pressure synthesis methods are established, the wear applications of coated BN tools are likely to be the initial commercial products.

Until now, however, attempts to synthesize cubic BN at low pressures have not been as successful as those to produce low-pressure diamond. Major basic problems have apparently not yet been satisfactorily solved (Rödhamer 1989). A direct analogy to the low-pressure diamond CVD synthesis appears to be an obvious possibility, however.

The recent annual market growth rates of superhard products were high and are projected to increase during the next couple of years. Superhard BN films and coatings should rapidly gain significant industrial importance, as BN exhibits an excellent inertness to hot steel surfaces under conditions where diamond is violently dissolved and therefore cannot be used. As known from the ultrahigh-pressure PCBN, this fact alone promises extraordinary possibilities for tools for shaping steels. In view of this potential market for highly relevant commercial applications, more and more research groups are working on the low-pressure CVD synthesis of superhard BN (Rödhamer 1989).

Besides the synthesis of the pure BN phases, the combined synthesis of diamond and BN could lead to new superhard materials with interesting properties complementing those of pure diamond and pure BN (Lux 1986; Badzian 1988).

5. General conclusions

Wear and cutting applications mainly utilize the superhard properties of diamond. Today, product development for these applications has already reached a relatively advanced stage, since single crystallinity and a perfect diamond crystal lattice are not primary requirements for their proper functioning.

Though their technical feasibility has been demonstrated and has already led to some industrial products, the production of technically and economically feasible products for superhard applications is still only in its early stages. The rapid advancement in these 'simpler' application areas will most certainly continue and even gain momentum.

Beside improving problems of economy, major technical problems, such as *scaling up* remain unsolved. A number of individual, as yet unpublished, larger-scale production setups are now being simultaneously developed. The selection of the 'best' method, to determine the construction of the most suitable equipment for specific products, is the big challenge at present. The CVD reactor for normal hard coatings is traditionally a batch-type unit. It seems that the initial *in situ* coating unit for low-pressure diamond coatings will be the same for both *in situ* and freestanding diamond products.

Though successful performance tests are encouraging and prove that the diamond coating adhesion for *in situ* coatings and freestanding diamond layers is not an insoluble problem, better adherence and general reproducibility are needed. Selection of the appropriate substrate chemistry or effective multilayer combinations can help to reduce interfacial stress.

Methods for a well controlled diamond nucleation, combined with reproducible substrate surface preparations, will help to achieve this goal.

Other important topics are composite powders as well as compacted products from such powders. Their future depends largely on their properties, on new applications and on their production costs.

Syntheses of superhard boron nitrides or other new superhard phases are still in the basic research stage. Both are an exciting challenge for the near future, especially in view of the considerable market potential of these materials for wear applications.

References

Badzian, A. R. 1988 *Advances in X-ray analysis*, vol. 31, pp. 113–127. New York: Plenum.

Badzian, A. R., Bachmann, P. K., Hartnett, T., Badzian, T. & Messier, R. 1987 In *Proc. E-MRS Meeting*, vol. XV, pp. 63–77. Les Editions de Paris.

Bichler, R., Peng, J., Haubner, R. & Lux, B. 1988 *Mater. Sci. Engng* A **105**, 543–547.

Genasystem Worthington, Ohio; Materials and Processing Report, vol. 5, No. 4/5 Elsevier.

Haubner, R. 1990 *J. Ref. Met. Hard Mat.* **9**, 70–77.

Heath, P. J. 1989 *VDI Berichte* **762**, 37–59.

Holleck, H. 1986 *Z. f. Werkstoffk.* **17**, 334–341.

Johannsen, P. 1989 *VDI Berichte* **762**, 259–271.

Kikuchi, N. 1989 *Diamond and DLC coatings*. Marco Island, Florida: Gorham Advanced Materials Institute.

Lindlbauer, A. 1992 Doctoral thesis, TU-Vienna, Austria.

Litos, R. 1992 Doctoral thesis, TU-Vienna, Austria.

Lux, B. 1986 European Patent, EP 0226898 A2.

Lux, B., Funk, R., Schachner, H. & Triquet, C. 1971 CH Expose d'Invention, 540990.

Lux, B. & Schachner, H. 1978 *High Temp. High Press.* **10**, 315–323.

Lux, B., Haubner, R., Altena, H. & Colombier, C. 1987 Fischmeister/Jehn. *Hartstoffschichten zur Verschleissminderung*, pp. 71–112. Oberursel, Germany: Verlag.

Lux, B., Haubner, R. & Wohlrab, C. 1989 *Surf. Coatings Technol.* **38**, 267–280.

Lux, B. & Haubner, R. 1989 *RHM* **8**, 158–174.

Lux, B. & Haubner, R. 1991 Diamond and diamond-like films and coatings. In *Proc. NATO-Advanced Study Inst. on Diamond and Diamond-Like Films and Coatings* (NATO-ASI Series B: Physics), vol. 266. New York: Plenum.

Lux, B. & Haubner, R. 1992 In *Diamond films and coatings* (ed. R. Davies). New Jersey, U.S.A.: Noyes.

Oakes, J., Pan, X. X., Haubner, R. & Lux, B. 1990 In *Proc. 1st Euro. Conf. on Diamond and Diamond-like Carbon Coatings* (ed. A. Matthews & P. K. Bachmann), pp. 600–607. Crans-Montana, Switzerland.

Patterson, D., Clin, J., Bai, B., Xiao, Z., Tomplin, N., Margrave, J. L. & Hange, R. 1991 In *Proc. 179th Meeting of the Electrochemical Society*. Washington, D.C.: U.S.A.

Reineck, I., Söderberg, S., Westergren, K. & Shahani, H. 1990 In *Proc. 2nd Int. Conf. on the New Diamond Science and Technology* (ed. R. Messier, J. T. Glass, J. E. Butler & R. Roy), pp. 809–814. Washington, D.C., U.S.A.

Rödhamer, P. 1989 In *Proc. 12th Int. Plansee-Seminar* (ed. H. Bildstein & H. M. Ortner), vol. 3, pp. 661–667. Reutte, Austria.

Rudder, R. A., Thomas, R. E. & Markunas, R. J. 1991 In *Proc. 179th Meeting of the Electrochemical Society*. Washington, D.C., U.S.A.

Sato, Y. 1990 In *Proc. Workshop on the Science and Technology of Diamond Thin Films*. Cleveland, Ohio.

Schachner, H., Tippmann, H., Lux, B., Stjernberg, K. & Thelin, A. 1984 Swedish Patent 8403428.

Schweitzer, R. 1989 *VDI Berichte* **762**, 293–303.

Sumitomo Electric Industries 1990 TAB bonding tool, EV-32R (August 1990) (commercial documentation).

Tanabe, K. & Fujimori, N. 1990 Japanese Patent Application, Tokugan-hei 2(1990)-81777.

Yashiki, T., Nakamura, T., Fujimori, N. & Nakai, T. 1991 In *Proc. Int. Conf. on Metallurgical Coatings and Thin Films*. San Diego, California.

12

Thermal and optical applications of thin film diamond

BY M. SEAL

Virtually all recent reviews of the market potential for chemical vapour deposited (CVD) diamond have featured the thermal management of electronic semiconductor devices as an imminent application for this new material. There is an existing market for natural diamond substrates ('heat sinks') in sub-millimetre sizes, and their thermal performance has been extensively studied. CVD diamond heat sinks in millimetre and larger sizes are already in use, but there are constraints to their applicability arising from thermal and mechanical factors. Their advantages and limitations are discussed.

The first 'optical' applications of CVD diamond films were as X-ray transmissive components (lithography masks and windows for soft X-ray detectors), but with improvements in the technology of CVD diamond growth a larger market for wide-band infrared transmissive windows is now developing. This results from the availability of large area (greater than 1000 mm^2) CVD diamond plates of adequate thickness and with transparency achieved through control of diamond grain size and orientation.

---

1. Introduction

Diamond has many extreme properties. They include hardness, strength, elastic moduli, gram atom number density (Angus 1986), thermal conductivity at room temperature, and spectral range of light transmission. This paper is concerned with the last two of these, and with the applications of thin film diamond that are based on them and likely to become important in the near to mid-term future. Those that depend directly on the thermal and optical properties of thin film diamond will most probably be, in the first place, extensions of similar applications of natural diamond.

Of the many uses of natural diamond where its high thermal conductivity plays a role, as abrasives, turning tools, or wire drawing dies for example, the use as a thermally conducting substrate (or 'heat sink') is probably the one which depends most directly on this property. It is one which has attracted the attention of virtually all analysts of the new market for chemical vapour deposited (CVD) diamond, perhaps because the much larger plate areas now becoming available in thin film diamond offer intriguing possibilities of extending an established market. The technical basis of these possibilities is discussed below.

The major use of diamond with relevance to the optical properties is of course the use as a gem. Its attractiveness in jewellery depends on the high refractive index and the high optical dispersion which yields a play of colour, plus the purity of its whiteness in the non-internally reflecting condition (or in some cases the subtlety of a fancy colour). However, gem diamonds are by their nature non-thin. They thus fall by definition outside the scope of this paper and it seems in any case highly unlikely

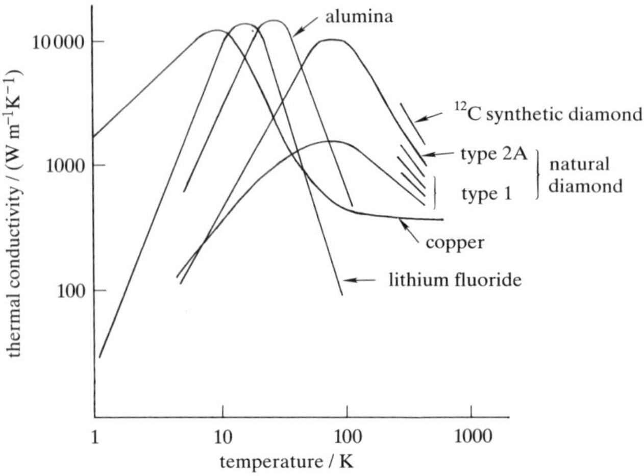

Figure 1. Thermal conductivities of various materials. Data from Berman (1965, 1973), Burgemeister (1978), Onn *et al.* (1992).

that growth of sufficiently thick, sufficiently pure plates of single crystal CVD diamond for gem application will be feasible and economic in the foreseeable future.

A more imminent optical application is the use as infrared transmissive windows. Diamond has a uniquely wide range of transparency, ranging in the case of high purity type 2A diamonds from the ultraviolet cut-off near 230 nm wavelength in the ultraviolet to beyond 100 μm in the far infrared, interrupted only by a relatively weak multi-phonon absorption system in the 2–6 μm region. Large flat CVD diamond plates are now available. They are polycrystalline, but their optical quality is improving rapidly with continued industrial development and they are by now effectively transparent through most of the infrared.

2. Thermal conductivity

It is probably still not generally realized that diamond is the best conductor of heat known at ambient temperatures. A good (type 2A) natural diamond conducts heat some five times better than copper at 300 K with a conductivity about 2000 W $m^{-1}K^{-1}$. As was pointed out by Berman *et al.* (1953) and Berman (1965, 1973) there is nothing surprising in this. It is a consequence of the tightly packed atoms and strong bonds of the diamond lattice, generally interpreted as resulting in easy phonon flow, coupled with a high Debye temperature. The functional dependence of conductivity on temperature is normal (figure 1), but with the curve shifted towards higher temperatures as compared with other dielectric crystals. Because impurity atoms scatter phonons, the thermal conductivity of diamond depends on its purity. Correlations have been established between conductivity and infrared spectral absorption features attributed to nitrogen in diamond (Schorr 1969; Martinez 1976; Burgemeister 1978). These are some of the spectral features used in assigning diamonds to type 1 or type 2 according to the physical classification first put forward by Robertson *et al.* (1934) and later sub-divided. Extreme type 1 (nitrogen-containing) diamonds are little better thermally than copper (Burgemeister 1978).

Above these there are all gradations in thermal conductivity up to the high values of the pure type 2A (figure 1).

Synthetic diamonds produced at high pressures and temperatures also show variation in thermal conductivity. At best they are comparable with natural type 2A (Slack 1973). The curves are similar in shape with absolute values reflecting the nitrogen content. However, the nitrogen is normally in non-aggregated (substitutional) form which has a smaller effect on thermal conductivity due to less effective phonon scattering (proportional to the product of nitrogen concentration and aggregate size). Also the amounts of nitrogen are lower, normally a few hundred parts per million (p.p.m.), as compared with the quantities of 1000 p.p.m. upwards which are common in natural diamond. There is also the possibility of removing the nitrogen from synthetic diamond by gettering the metal solvent during growth. The result is a diamond thermal conductivity which is normally in the range 1200–2000 W m^{-1} K^{-1} at 300 K and for gettered samples near 2000 W m^{-1} K^{-1}. One may also increase the thermal conductivity of synthetic diamond by making it from isotopically purer carbon, as compared with carbon of the natural isotopic abundance (1.1 % ^{13}C). Anthony *et al.* (1990) grew diamond by a two-stage CVD/high pressure process from carbon containing only 0.1 % ^{13}C and obtained a thermal conductivity (deduced from the diffusivity) of 3300 W $m^{-1}K^{-1}$.

Thermally CVD diamond has the advantage that it can be grown reasonably pure chemically, though some processes are more prone to give contamination than others. The hot filament method is, for instance, likely to introduce the filament metal as an impurity in the sample. Also the carbon is not always incorporated with sp^3 bonding. Regions with sp^2 bonding do occur. However, the main problem thermally is the polycrystalline nature of all the CVD diamond films that have so far been considered for thermal applications. The grain size increases as the film grows thicker and this results in a variation of thermal conductivity through the layer (Graebner *et al.* 1992). These authors reported a thermal conductivity of at least 2100 W $m^{-1}K^{-1}$ at the growth surface of a 350 μm diamond film, but only about 7 W $m^{-1}K^{-1}$ near the substrate interface.

Polycrystalline CVD diamond films show an anisotropy of thermal conductivity between directions parallel to and perpendicular to the film plane (Graebner *et al.* 1992). That perpendicular to the plane was found to be at least 50 % higher than that parallel to the plane. There is variation with growth conditions and for a sub-set of these the conductivity has been found to vary inversely with the growth rate and Raman line width (Graebner *et al.* 1992). Overall variation over at least a range 200–2100 W $m^{-1}K^{-1}$ at 300 K has been reported in the literature for a range of CVD diamond samples. Values for currently available commercial films seem to be about 1100 W $m^{-1}K^{-1}$ or possibly somewhat higher (Visser *et al.* 1992).

3. Heat sinks

(*a*) *Type 2A diamond*

The optimum dimensions and proportions of natural diamond heat sinks have been extensively investigated both experimentally and theoretically. The experimental investigations have been complicated by thermal resistances in the semiconductor chip materials, in solder layers, and at interfaces, but appear in all cases to be consistent with theory. The definitive theoretical study is probably that of Molenaar & Staarink (1985), extended later by Doting & Molenaar (1988). These

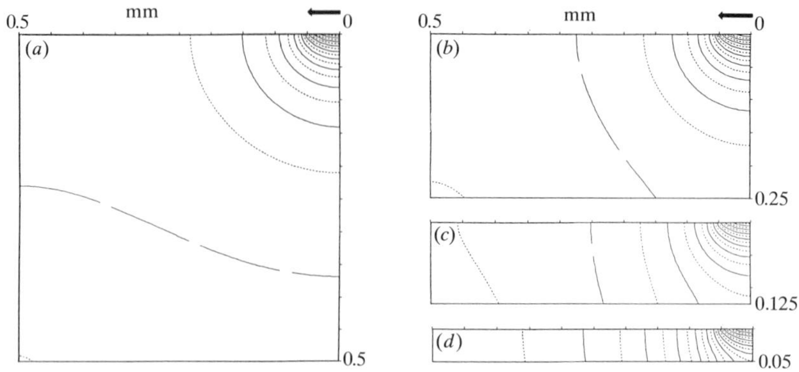

Figure 2. Isotherms in half-sections of diamond heat sinks. See text for details. Courtesy Drukker International B.V.

authors based their calculations on Burgemeister's (1978) result for type 2A diamond that the thermal conductivity (in MW $m^{-1}K^{-1}$) over a temperature range 300–450 K can be represented by

$$k(T) = 2.75T^{-1.26}. \tag{1}$$

They considered various geometries including diamond heat sinks mounted on copper (mesa mounting) and embedded in copper (flush mounting). The semiconductor chip was assumed centrally mounted on the diamond and both square and circular axial symmetries were treated for the mesa-mounted case (cylindrical only for flush mounting). These geometries represent those used in the mounting of microwave diode chips. The copper mount was assumed plane-bounded semi-infinite. In practical cases it is generally large enough that this assumption will not introduce excessive error. The copper was taken to have a thermal conductivity of 387 W $m^{-1}K^{-1}$, not varying with temperature. A geometry approximating that used for mounting semiconductor diode laser chips was also studied. This has an elongated rectangular 'stripe-geometry' chip edge-mounted on an assumed quarter-infinite copper block (defined by two perpendicular plane surfaces). The stripe is mounted on one surface with its length perpendicular to the edge.

The mathematical treatment proceeded from the heat flow equation (conservation of energy in an elemental volume) directly to a Laplace equation in the copper and via a derived function in the diamond. After definition of boundary heat fluxes a computer iterative method yielded a converging solution for the distribution of isotherms in the diamond heat sink. This method started with an arbitrarily assumed homogeneous heat flux across the diamond–copper boundary, calculated from the copper side the temperature distribution over this boundary, computed the derived function of temperature everywhere in the diamond including all boundary conditions, and calculated a new temperature distribution over the diamond–copper boundary. Combination of new and old distributions with a weighting factor for rapid convergence yielded after appropriate iteration a self-consistent solution.

Though differing in detail, the solutions for the different mesa-mounted geometries showed many similarities. The proportions, and in particular the aspect ratios of diamond thickness to lateral dimensions, are extremely important. Figure 2 shows isotherms for a representative series of geometries. The plots each represent a half

Table 1.

diamond thickness/µm	maximum temperature rise/°C
500	121
250	117
125	119
50	141

section of a mesa-mounted cylindrical heat sink with the axis at the right. The chip is mounted centrally at the top right, the arrow indicating its radius. The lower surface is mounted on copper, which extends to a large distance on either side. The cylinder diameter is 1 mm and the chip diameter 0.1 mm. The chip is assumed to dissipate 20 W uniformly over its lower surface. The isotherm interval is 5 °C. The effect of progressive reduction in diamond thickness is shown in the series figure 2*a*–*d* (0.5–0.05 mm). Table 1 lists the corresponding maximum temperature rises centrally under the chip. If one starts from the best geometry (*b*) it is clear that doubling the thickness of diamond from 250 µm to 500 µm has resulted in a slight worsening of performance, a disappointing result for a substantially more expensive heat sink, whereas reducing the thickness by a factor of 2 gives only marginal worsening. Reduction by a factor 5 gives substantial worsening.

The reasons are clear from figure 2. The top half of (*a*) shows roughly hemispherical isotherms, i.e. good heat spreading. The lower half shows something like linear heat flow through a bar. It is an unnecessary thermal resistance not contributing significantly to the spreading. At the other extreme (*d*) there is little spreading and a high temperature rise (about 85 °C) at the centre of the diamond–copper interface. In case (*a*) there is a major heat flux through the outer part of the diamond–copper interface (particularly near the outside edge where the cooling effect is enhanced by that of the copper outside the diamond edge). In case (*d*) most of the heat flow is through the central part of the diamond–copper interface and the spreading resistance of the copper from that central part becomes dominant. The isotherms in the outer part of the diamond in (*d*) show some lateral heat transport in the diamond layer, but the flux is low because of the small cross section.

Clearly the diamond can be too thick or too thin. Experience has led to a rule of thumb that ideally the thickness should be about one-third of the lateral dimensions with a range from about one-quarter to one-half acceptable. Down to one-tenth may sometimes be acceptable. In turn the lateral dimensions should be related to the chip dimensions. The ratio of ten used in figure 2 is larger than normal. It was chosen to illustrate the effects better. A ratio of five would be more normal. Similar considerations apply in the case of the edge-mounted, stripe-geometry, laser diode chip. The diamond square size should be at least a factor of 2 larger than the stripe length and preferably a factor of 3. The thickness should be about equal to the stripe length, though thicknesses down to about one-quarter of the stripe length may be acceptable with loss of efficiency. There is a wider range of acceptable thicknesses on the high thickness side, but little point in using this. In summary for mesa-mounted chips, the diamond heat sink lateral dimensions should be about 5 times the chip dimension for centrally mounted chips and $2\frac{1}{2}$ times for edge mounted chips corresponding to the lower solid angle of heat flow. The diamond thickness should be about one-third of the lateral dimensions in either case, though with more flexibility in the case of the edge-mounted chip (because of the thinness of the stripe).

Embedded diamond heat sinks behave quite differently, however. Since diamond replaces copper, a larger diamond is always better than a smaller, though the gain in advantage factor becomes very small beyond diamond diameters about 10 times the chip diameter or diamond thicknesses twice the diamond diameter. On the low thickness side there is never likely to be much point in using diamonds thinner than about one-tenth of properly chosen lateral dimensions.

(*b*) CVD *diamond*

As indicated in §2 above, the thermal conductivity of CVD diamond differs from that of natural in that it may vary through the thickness of a sample and be anisotropic. Absolute values may vary over a wide range depending on the manufacturing process and the exponent in equation (1) may vary. The Molenaar-Staarink treatment of §3*a* is thus not directly applicable, though changes of uniform conductivity values and exponent should be trivial. A tensor treatment for the anisotropic case might be possible, but is certainly far from trivial. However, some general conclusions look possible. Since the diamond layers are heat spreaders it is the lateral conductivity which is limiting. In an extreme case of diamond film having columnar crystallites of excellent conductivity isolated thermally from each other, the crystallites would simply channel the heat downwards. There would be no heat spreading and the diamond layer would be useless thermally. Fortunately it is the limiting lateral thermal conductivity that is usually measured and materials with high values of this are obtainable.

The most obvious advantage of CVD diamond films is that they can be of relatively large area, an order of magnitude larger in linear dimensions than is normally possible with single crystal diamond. This brings two principal claimed advantages. One is that wafer processing of small components should be possible, provided the problems of polishing and dicing diamond wafers can be solved. The other is that larger diamond components should become economic and thus open up new thermal management applications in electronics for diamond. In the wafer processing case the products, if made from the highest conductivity CVD material, should be capable of directly replacing natural diamond heat sinks. The advantages will be in price and ease of applying metallization patterns by photolithographic means. The second is much more problematic.

If the 'one-third lateral dimensions' rule of §3*a* holds, or even a one-tenth lateral dimensions rule, available thicknesses of CVD diamond are going to need to increase as component sizes increase. The differences between natural and CVD diamond are likely to increase the critical thickness. One may thus assume that CVD diamond heat sinks for 3 mm square transistor chips, which would need to be 15 mm square and 1.5–5 mm thick, are not yet near commercial reality. Perhaps they will become so as people learn to make thicker diamond films more cheaply, but at the moment such thicknesses look to be prohibitively expensive.

There is, however, a possibility for use of larger area CVD diamond heat sinks in the thermal management of devices having multiple small heat sources on a larger chip. An example would be a laser diode array chip having perhaps 10 well-separated, light-emitting stripes on a single chip. The small size of the individual stripes determines the critical thickness of the substrate, whereas a larger area of substrate is necessary to accommodate the rather large chip. Large area CVD diamond heat sinks may also be appropriate in cases where a hard, mechanically stable substrate is needed simply to conduct heat rather than spread it. This might be the case where

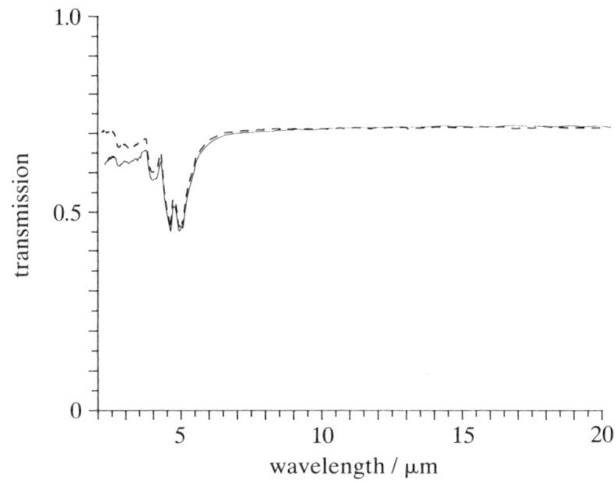

Figure 3. Infrared transmission spectra of a natural type 2A diamond and a CVD diamond film. Courtesy Norton Company. ——, TD826A; ---, natural.

active cooling is involved (micro-channel fluid cooling, heat pipes, Peltier cooling elements).

4. Optical applications

(*a*) *Infrared windows*

Optical applications obviously depend on the availability of optical quality CVD diamond. For the foreseeable future this is likely to have to be polycrystalline material and there is the problem of light scattering at grain boundaries rendering the material translucent or opaque. The present situation seems to be that the best material is only translucent in the visible, but effectively transparent in the infrared because of the longer wavelength. Figure 3 shows an infrared transmission spectrum of a sample of currently commercially available material. The transmission is equivalent to that of natural type 2A diamond at the longer wavelengths and starts to fall off towards shorter wavelengths. It is limited to slightly over 70% because of Fresnel reflection losses at the high refractive index diamond–air interfaces.

CVD diamond infrared transmissive windows have major potential military markets. The combination of strength, hardness, chemical inertness, and high thermal conductivity with good optical properties would make them for instance particularly suited for use as the outer windows on infrared missile seeker devices flying at high Mach numbers. In the civilian sphere there is increasing interest in infrared instrumentation for industrial process control. Kilham & Le Blon (1990), for instance, describe a flow cell using diamond windows for the on-line analysis by infrared spectroscopy of molten polymers in the plastics manufacturing industry. They give as an example the measurement of concentrations of carbonyl groups resulting from oxidative degradation upstream in the continuous flow production of polyethylene. Real time detection of such degradation can prevent the manufacture of tonnage quantities of defective product (which might result if off-line analyses only were used). The cost savings are obvious. Such techniques have been used with sapphire windows in the near infrared (e.g. for the detection of water) but diamond transmits in spectral regions in the mid infrared (where sapphire does not) and these are of importance for the detection of organic groups.

Natural diamond windows have also been used in research projects involving infrared spectroscopy. The most spectacular of these was an 18.2 mm diameter window used in the Pioneer Large Probe experiment to investigate radiation energy balance and spectral distribution from measurements inside the Venusian atmosphere (Boese *et al.* 1979). There have also been many more down-to-earth projects as well as routine applications in instrumentation. Some have been reviewed by Seal & van Enckevort (1988). Besides spectroscopic applications natural diamond windows have also been used as simple, light-transmissive elements separating for example body fluids from instrument space in medical endoscope or laser therapy applications, or as exit windows for CO_2 lasers (diamond being particularly appropriate for transmission of their 10.6 μm radiation).

The vast majority of optical applications require physical apertures greater than can be met from natural diamond. CVD diamond has the potential to remove this restriction. If optical transmission quality and finishes can be produced, there will be the potential for a very large business as diamond has some most attractive properties. It is hard and scratch resistant. It is inert to virtually all the chemical environments that a window is likely to meet. It is strong, a fault free diamond window of 0.25 mm thickness being adequate to withstand a 1 atm ($\approx 10^5$ Pa) pressure differential to quite large diameters (say 25–100 mm depending on the permissible central deflection, method of mounting, and safety factor required). The spectral transmission range is enormous. Even the C–C absorption systems between 2–6 μm are weak (figure 3) to negligible in thin windows below say 50 μm thickness. On the negative side there would be problems of chromatic aberration in refractive components and the reflection is high (about 17% at a single interface). Anti-reflection coatings may thus be required for some applications.

(*b*) *Other optical applications*

One of the original applications of CVD diamond films was as X-ray transmissive components. The low atomic number ensures good X-ray transmission. High optical transparency, high strength, high thermal conductivity, low thermal expansion, and excellent chemical inertness are processing aids or requirements for many X-ray applications. The chief of these were as X-ray lithography masks and transmissive windows for soft X-rays. Thicknesses are low – a few micrometres at most – and support is a problem. Support grids can be used in the soft X-ray window case, as for example in windows protecting light element detectors in X-ray analysis adjuncts to scanning electron microscopy such as EDAX. Supports in the lithography case clearly must not interfere with the pattern and tensioned films are needed. Most of the details are subject to industrial secrecy, but Löchel *et al.* (1992) have recently surveyed the rather fragmentary literature. Other earlier papers are by Cuomo *et al.* (1991) and Windischmann *et al.* (1991). Additional possibilities exist of active optoelectronic uses. These are considered to fall outside the scope of the present paper, but might include ultraviolet and other radiation detectors, light emitting diodes, and possibly short wavelength lasers based on radiation-induced H3 centres (Rand & DeShazer 1985).

5. Conclusions

The principal thermal and optical applications of CVD thin film diamond that are likely to be of substantial commercial importance in the short term are as heat spreading substrates in electronics (heat sinks) and as infrared transmissive windows.

Both are extensions of current markets for natural diamond, based on lower costs and larger areas. Diamond membranes for X-ray masks will also be important and thin diamond windows for soft X-ray detectors fill a needed niche.

References

Angus, J. C. 1986 Empirical categorization and naming of 'diamond-like' carbon films. *Thin Solid Films* **142**, 145–151.

Anthony, T. R. *et al.* 1990 Thermal diffusivity of isotopically enriched ^{12}C diamond. *Phys. Rev.* B **42**, 1104–1111.

Berman, R. 1965 Thermal properties. In *Physical properties of diamond* (ed. R. Berman), pp. 371–393. Oxford: Clarendon.

Berman, R. 1973 Heat conductivity of non-metallic crystals. *Contemp. Phys.* **14**, 101–117.

Berman, R., Simon, F. E. & Ziman, J. M. 1953 The thermal conductivity of diamond at low temperatures. *Proc. R. Soc. Lond.* A **220**, 171–183.

Boese, R. W., Pollack, J. B. & Silvaggio, P. M. 1979 First results from the large probe infrared radiometer experiment. *Science, Wash.* **203**, 797–800.

Burgemeister, E. A. 1978 Thermal conductivity of natural diamond between 320 and 450 K. *Physica* B **93**, 165–179.

Cuomo, J. J., Doyle, J. P., Pappas, D. L., Saenger, K. L., Guarnieri, C. R. & Whitehair, S. J. 1991 High density amorphous carbon films and the preparation of diamond membranes for X-ray lithography. In *Applications of diamond films and related materials* (ed. Y. Tzeng, M. Yoshikawa, M. Murakawa & A. Feldman), pp. 169–180. Amsterdam: Elsevier.

Doting, J. & Molenaar, J. 1988 Isotherms in diamond heat sinks, non-linear heat transfer in an excellent heat conductor. In *Proc. 4th SEMI-THERM*. Piscataway, New Jersey: IEEE.

Graebner, J. E., Jin, S., Kammlott, G. W., Bacon, B., Seibles, L. & Banholzer, W. 1992. In Anisotropic thermal conductivity in CVD diamond. *J. appl. Phys.* **71**, 5353–5356.

Graebner, J. E., Jin, S., Kammlott, G. W., Herb, J. A. & Gardinier, C. F. 1992 Unusually high thermal conductivity in diamond films. *Appl. Phys. Lett.* **60**, 1576–1578.

Graebner, J. E., Mucha, J. A., Seibles, L. & Kammlott, G. W. 1992 The thermal conductivity of chemical-vapor-deposited diamond films on silicon. *J. appl. Phys.* **71**, 3143–3146.

Kilham, L. B. & Le Blon, M. W. 1990 Sampling flow cell with diamond window. U.S. Patent 4,910,403.

Löchel, B. *et al.* 1992. Diamond membranes for X-ray masks. *Microelectron. Engng* **17**, 175–180.

Martinez, M. 1976 D.Phil. Thesis, Oxford University, U.K.

Molenaar, J. & Staarink, G. W. M. 1985 The optimal form of diamond heat sinks. In *Proc. First Eur. Symp. on Mathematics in Industry* (ed. M. Hazewinkel, R. M. M. Mattheij & E. W. C. van Groesen), pp. 113–126. Stuttgart: Teubner.

Onn, D. G., Witek, A., Qiu, Y. Z., Anthony, T. R. & Banholzer, W. F. 1992 Some aspects of the thermal conductivity of isotopically enriched diamond single crystals. *Phys. Rev. Lett.* **68**, 2806–2809.

Rand, S. C. & DeShazer, L. G. 1985 Visible color-center laser in diamond. *Optics Lett.* **10**, 481–483.

Robertson, R., Fox, J. J. & Martin, A. E. 1934 Two types of diamond. *Phil. Trans. R. Soc. Lond.* A **232**, 463–535.

Schorr, A. J. 1969 A comprehensive study of diamond as a microwave device heat sink material. In *Proc. Industrial Diamond Conf.*, pp. 185–190. Moorestown, New Jersey: Industrial Diamond Association of America.

Seal, M. & van Enckevort, W. J. P. 1988 Applications of diamond in optics. *Proc. SPIE* **969**, 144–152.

Slack, G. A. 1973 Nonmetallic crystals with high thermal conductivity. *J. Phys. Chem. Solids* **34**, 321–335.

Visser, E. P., Versteegen, E. H. & van Enckevort, W. J. P. 1992 Measurement of thermal diffusion in thin films using a modulated laser technique: application to chemical-vapor-deposited diamond films. *J. appl. Phys.* **71**, 3238–3248.

Windischmann, H., Epps, G. F., Caesar, G. P. & Maluf, N. I. 1991 Properties of diamond membranes for X-ray lithography masks. In *New diamond science and technology* (ed. R. Messier, J. T. Glass, J. E. Butler & R. Roy), pp. 791–796. Pittsburgh, Pennsylvania: Materials Research Society.

Index

Page numbers appearing in **bold** refer to figures and those in *italic* refer to tables.